4.50-1
E225

Rivers and River Terraces

EDITED BY
G. H. DURY

PRAEGER PUBLISHERS
New York · Washington

BOOKS THAT MATTER

Published in the United States of America in 1970
by Praeger Publishers, Inc., 111 Fourth Avenue,
New York, N.Y. 10003

Selection and editorial matter Copyright 1970 in London,
England by G. H. Dury

All rights reserved

No part of this publication may be reproduced, stored
in a retrieval system or transmitted in any form or by
any means, electronic, mechanical, photocopying, recording
or otherwise, without the prior permission of the
Copyright owner.

Library of Congress Catalog Card Number: 78-124605

Printed in Great Britain

DEDICATED TO

LUNA B. LEOPOLD

AND

WALTER B. LANGBEIN

Contents

	ACKNOWLEDGEMENTS	9
	INTRODUCTION	11
1	Methods and Results of River Terracing BY *Hugh Miller*	19
2	River Terraces in New England BY *W. M. Davis*	36
3	Longitudinal Profiles of the Upper Towy Drainage System BY *O. T. Jones*	73
4	Land Sculpture in the Henry Mountains BY *G. K. Gilbert*	95
5	Erosional Development of Streams: Quantitative Physiographic Factors BY *R. E. Horton*	117
6	Flood Plains BY *M. Gordon Wolman and Luna B. Leopold*	166
7	River Channel Patterns BY *Luna B. Leopold and M. Gordon Wolman*	197
8	River Meanders and the Theory of Minimum Variance BY *Walter B. Langbein and Luna B. Leopold*	238
9	General Theory of Meandering Valleys and Underfit Streams BY *G. H. Dury*	264
	INDEX	277

Acknowledgements

'Methods and Results of River Terracing', by Hugh Miller © *Proc. Roy. Phys. Soc. Edinburgh*, VII (1883)

'River Terraces in New England', by W. M. Davis © *Bull. Museum Comparative Zoology*, XXXVIII (1902)

'The Longitudinal Profiles of the Upper Towy Drainage System', by O. T. Jones © *Quart. J. Geol. Soc. London*, LXXX (1924)

'Land Sculpture in the Henry Mountains', by G. K. Gilbert, 1887, U.S. Geographical and Geological Survey of the Rocky Mountain Region (now the Geological Survey)

'Erosional Development of Streams: Quantitative Physiography Factors', by R. E. Horton © *Bull. Geol. Soc. Amer.*, LVI (1945)

'Flood Plains', by M. Gordon Wolman and Luna B. Leopold, 1957, U.S. Geol. Survey, Prof. Paper 282-C

'River Channel Patterns', by Luna B. Leopold and M. Gordon Wolman, 1957, U.S. Geol. Survey, Prof. Paper 282-B

'River Meanders and the Theory of Minimum Variance', by Walter B. Langbein and Luna B. Leopold, 1966, U.S. Geol. Survey, Prof. Paper 442-H

'Meandering Valleys and Underfit Streams', by G. H. Dury, U.S. Geol. Survey, Prof. Paper 452-A-B-C

Introduction

ANTHOLOGISTS run the occupational risk of satisfying few readers, if indeed any; but they may perhaps find comfort in the fable of the father, the two sons and the ass. I am of course aware of very many papers, additional to those actually selected, which could justly have been included; and I am equally aware that the chosen list is short. The range of sampling could have been increased, and the list of items extended, had it been possible to deal entirely in terms of brief extracts; but then the nature of the material could have forced the collection into the undesired form of a monograph, a review article or a dissertation. This would have been to defeat the central purpose of assembling original material from a variety of sources, and of presenting it in conveniently abbreviated form but still at significant length.

As first published, the original works run in text and illustrations to a combined equivalent of more than a third of a million words. The permitted space in this book involves an average reduction to less than a quarter of the original wordage. Accordingly, in undertaking the ruthless task of selection I have been led to exclude works written in languages other than English. Conscious as I am of a massive debt to writers in French and German – among them for instance de Martonne, Troll, Macar, and Tricart – I can do no more than place on record my regret at the obvious omissions.

The chosen items all strike me as fundamental, in some sense or other, to the set theme of rivers and river terraces. They are also intended to illustrate some of the direction taken by studies of fluvial morphology generally. Were demonstration necessary, it could surely be proved that the works by Gilbert, Miller, Davis and Jones, first appearing in the interval 1877–1924, have proved highly influential on a number of subsequent writers; indeed, it can safely be claimed that some of their influence still persists in the minds and writings of those who, by direct means or indirect, have absorbed their ideas and their examples. Horton's paper stands in the direct line of descent from Gilbert's quantitative and experimental work, which is less well known among geomorphologists generally than is the earlier material used here. Furthermore, as scarcely needs to be pointed out, Horton's work is the immediate and direct precursor of numerous recent morphometric studies. The papers of Leopold and Wolman, Wolman and Leopold, and Langbein and Leopold exemplify the direction and results of investigations into streams and stream behaviour during the late 1950s and 1960s, illustrating the progress from enlightened empirical research through straightforward quantification to the use of stochastic methods. In selecting a brief summary of my own findings for inclusion in a series which I have called fundamental, I do not wish to seem lacking in modesty; the item is required for use in the critical evaluation of the papers by Miller and Davis.

Miller's paper constitutes a quite early, and in some ways a quite full, disquisitional inquiry into terraces in their relation to streams. It is widely cited in the literature, although seldom discussed; and one may speculate that it might not have become so widely known as it is, at least by title, had it not been the acknowledged forerunner of Davis's own account of the river terraces of New England. However, Miller's study is well worthy of attention for its own sake; in listing nine possible modes of origin for terraces, it agrees with the multicausational attitude now being adopted by a number of influential theoretical geomorphologists. It is highly commendable for its appeal to numerical data and to systematic and quantitative conclusions reached by other workers; it also deserves praise for its wide coverage of previous literature. The one point of caution to be observed in reading either the original or the condensed version offered here is that Miller is not wholly consistent in his use of the term 'terrace'; the context shows that, at a number of points, he intends to connote the fore-edge alone.

It would scarcely have been possible altogether to omit from this anthology the work of Davis. Beginning with the proposition that the most appropriate source for this author is *Geographical Essays*, I found myself proceeding as it were by elimination. Of the fourteen physiographic essays, no more than four or five can be regarded as central to the present theme; and I have already had occasion elsewhere to examine two of these in some detail. Neither of the pair entitled 'Rivers and Valleys in Pennsylvania' and 'The Rivers of Northern New Jersey' seems especially well adapted to the purpose in hand. The final choice thus falls on 'River Terraces in New England'. Although some portions of this have also been formerly criticised by me in a particular context, the version given below is designed as a discussion of terrace cutting rather than one of hypothetical changes in volume, and may therefore be made to stand on its own. On the positive side, I consider this essay to exemplify as well as any the Davisian method of treatment and style of writing. Its strengths lie in the rapid and repeated application of a suddenly conceived and illuminating idea to a whole series of field examples. Its weaknesses include its limited scope in comparison to Miller's study, its failure to appeal to measurement, and its paucity of site investigation. Estimates of the height of fore-edges (which, like Miller, Davis is capable of calling 'terraces') remain merely estimates. The survey mentioned as desirable in the full text was never carried out by Davis, and the augering suggested as a check on the presence of concealed bedrock spurs in numerous cusps was not performed until years later, and then by others. Had Davis made the check himself, he would have found his hypothesis of defended terraces less than fully competent to explain the observed forms.

Jones's paper, projected in part as an empirical investigation of profile characteristics in the upper part of a single catchment, belongs in its historical context to the sequence of works which deal with high-level platforms, and in which reliance is placed on large-scale vertical shifts of the strandline. It is probably true to state that Wales, more than any other area, has been used by workers in Britain to sustain the hypothesis of intermittent falls of the strandline from very high stands indeed. Unlike Miller and Davis, and still earlier Hitchcock, whose influence pervades Miller's paper, Jones was not dealing with the dissection of valley fills of glacial origin; instead, he separates the observed effects which he has under study from the erosive influence of local ice. He makes use of empirical field measurements of profile characteristics, and undertakes a mathematical analysis of them, long before field measurement was at all common in geomorphology, and long before the incoming of quantification in its present guise. His paper is ancestral to numerous subsequent works in which terraces are referred to former sea stands, and where attempts are made to reconstruct these stands by the extrapolation of profiles. In the usual manner, this method of attack has of course been superseded by more recent work; it is now well established that extrapolation of profiles can, at best, provide no more than uncertain fixes, while the results of later studies of channel slope and its controls show that slope is determined by factors additional, or alternative, to stream volume. Nevertheless this item remains in advance of its time.

Introduction

That portion of Gilbert's work which appears below illustrates the well-established geologic tradition of using an empirical paper as the vehicle of theoretical statement. Charged with the investigation of a vast piece of new country, Gilbert took full advantage of the openness of the landscape to observe structure–surface relationships. He realised that stream channels and stream nets cannot be adequately studied without reference to the entire surfaces of the catchments involved, right up to the very crests of the divides. In this regard his work foreshadows that of Horton, and through Horton the productive explorations of the Columbia school of geomorphology and its adherents. Independently of all this, Gilbert's writing merits inclusion as introducing original terms and concepts which have subsequently been absorbed into the general corpus of geomorphic thinking. To vary the figure, it can be said that some of them still flow in the main current of geomorphic debate; thus in the selected passages below there appears the term 'dynamic equilibrium', which at the time of writing appears good for years of definition, analysis, re-analysis and disputation, whether useful or not.

Some of the readers for whom the present text is meant may not find the selected works of Miller, Jones and Gilbert particularly accessible. Horton's original, while readily available, is uncomfortably long for the purposes of some readers. Accordingly, I trust that the still lengthy but considerably abbreviated version offered below will prove of service. Representing the outcome of studies of the kind initiated by Gilbert, this paper constitutes also the product of years of original thought and inquiry by Horton himself. If any single paper marks the beginning of modern quantitative geomorphology, it is this one. The fact that the rapid advances of the last twenty years have reduced it in part to a work of historical interest is at the same time a tribute to its effectiveness, and a justification of its inclusion. Its implications have probably still to be fully worked out. The one important qualification which needs to be borne in mind in reading it is that, as Strahler has shown, there are advantages in revising Horton's stream-ordering system in the first and second orders, allocating to the second order a stream formed by the union of two fingertip tributaries.

It is with a certain diffidence that I have continued the sequence with the papers by Wolman and Leopold and by Leopold and Wolman, since their conclusions are incorporated in the well-known 1964 textbook *Fluvial Processes in Geomorphology* of Leopold, Wolman and Miller. However, there is room here for a fuller presentation of the relevant material than is provided in a more general work, and the two items lead directly on to important later research of a somewhat different kind, here represented by the 1966 paper of Langbein and Leopold. As is well demonstrated by the three papers taken together, these workers and their associates have shifted their main attention from empirical studies in the field or the laboratory to the arrival at general conclusions and the development of general theory. This statement is by no means intended as adverse criticism – indeed, precisely the reverse. Furthermore, it is in no way meant to imply that any of the workers in question tend to neglect observation; on the contrary, each piece of investigation begins, just as Gilbert's flume experiments earlier began, as a field or laboratory problem, and, as the selected items fully show, the combination of observation, experiment and analysis is impressive throughout. At the same time it seems reasonable to draw attention to developments in method of attack, wherein these three papers epitomise much of the recent conceptual growth of fluvial morphology.

The paper on flood plains combines quantitative observation of rates of deposition with an application of magnitude–frequency relationships. That on channel patterns deals with channel characteristics and with the complex interaction of the hydraulic variables which control these characteristics and which produce identifiable patterns in plan. Like the first paper, this too is deeply concerned with process, but it enters the realm of general theory when it relates identifiable patterns of channel to identifiable states of quasi-equilibrium. These two papers have brought into fluvial morphology much that is new, and still more that is illuminating. Their empirical content can scarcely fail to stimulate a great deal of further empirical study. However, such study is of full value only when it produces new ideas, or at least when it constitutes check experiments; as Chorley has pointed out, imitative *ad hoc* studies are markedly restricted in worth. The desirable next development is to further the search for general theory, as is done by Langbein and Leopold in the third of the papers under discussion. Although this work begins with the observation of actual meander patterns, this fact merely constitutes the connection of the theoretical exploration with the real world. In obtaining the sine-generated curve as the most suitable form to describe a meandering trace, the two authors employ a random-walk model. In showing that a meandering trace ensures minimum variance among hydraulic parameters, as in stressing steady-state conditions, they are conducting an inquiry into the operation of stochastic processes. The paper illustrates how far things have come since Horton's day. Quantification in the sense of numerical measurement and data processing of the simpler kind is being outrun by sophisticated calculations of probability and by the construction of new conceptual models. Development in this particular direction is so swift that its nature and pace are perhaps not yet fully appreciated or even comprehended. It can be urged that fluvial morphology in the last twenty years has already effected two internal revolutions and is embarked on a third. It has embraced numerical morphometry; it has assimilated the statistical treatment of a whole range of data on stream channels and of the interrelationships of these data; and it is now producing models with a stochastic base.

In preparing the shortened versions of the several papers chosen, I have needed to exercise a good deal of editorial discretion, quite apart from that involved in making the initial choice of items. As I have remarked, the total length of material has been greatly reduced. Compression of technical material can probably be more readily defended than can compression of writing meant originally as literature. In literature, form and content go inseparably together at a length properly determined by the author, whereas abstracts and summaries are routine in technical writing. Here, content is all, or nearly all, that has been eliminated, for I trust that contrasts of style are reflected in the chapters below.

It would have been possible, although hopelessly clumsy, to effect abbreviation by using the customary rows of dots to mark omissions, and square brackets to enclose words and phrases added for the sake of continuity. Instead, although I have in fact worked mainly by deletion, I have cast the reduced versions in the form of continuous writing. It follows that this book is not suitable for use as a primary source, even though it is meant to deal justly with the statements and views of the original authors. In the same connection the chapter headings have been altered, or at least rearranged, into a form suited to the purpose in hand.

Deletion has been made to fall, wherever possible, on whole sections or whole paragraphs, or at least on whole sentences, rather than on individual words and phrases. Many of the original diagrams and all of the original photographic illustrations have been omitted. The retained diagrams have been renumbered; all have been redrawn or at least relettered, in order to ensure a sensible uniformity of

Introduction 17

illustrative style; and some have been consolidated or rearranged, in order to give the maximum convenient size of drawing within the limits imposed by size of frame.

The various authors have fared differently among themselves during the process of condensation. From Miller's paper, discursive and parenthetical material and numbers of repetitive examples have been taken out. The lengthy footnotes of the original have been removed, any retained material being incorporated in the running text or transferred to the consolidated list of references. In this list, Miller's quite numerous omissions of dates, authors' initials and page references have been made good; I am indebted for help in the task of restoration to the librarians of the Royal Society of Edinburgh, the Geological Society of London and the United States Geological Survey.

As with Miller, so with Davis, the editorial process has fallen unevenly on the work as a whole. Certain complete paragraphs, and shorter passages which contribute little to the argument, have been struck out, while superfluous phrases have been removed from the retained material at very many points. In view of my disapprobation of Davis's writing it seems no more than reasonable to state that any stylistic defects which make themselves apparent in Chapter 2 are those of the original author: I refer in particular to the recurrent use of catch-phrases, instead of the employment of an emotionless technical vocabulary.

Jones's work is represented by what is, in the main, a verbatim but slightly shortened extract. Similarly, Gilbert's writing has not been much altered, except for some slight modernisation of vocabulary, phrasing and punctuation; the relevant item is again essentially an extract, although the section on equal and unequal declivities has been transferred from its original location in the monograph to the discussion of badlands, where, considering the design of the complete entry, it properly belongs. Horton's article, again of monograph length in the original, proved in some ways the most difficult of all to shorten to the obligatory length. In the event, it has been reduced mainly by block omissions, the text resisting most efforts at minor abbreviation. The retained portions, still quite bulky in total, are those which seem most directly and most inescapably relevant to the theme set.

The reduced versions of the three U.S. Geological Survey Professional Papers in which Wolman, Leopold and Langbein are concerned do not include those portions of the originals which, on a somewhat strict assessment, could be regarded as slightly less than centrally significant to the main arguments pursued. Here again the task of compression has not been an easy one, but it has, however, been lightened by the omission of all photographic illustrations and of text referring to them. In preparing my own contribution I have felt able to undertake draconian measures, running to little greater length than the total abstract length of the initial published works. The one liberty I have taken here is to introduce by name a type of underfit stream which remained still nameless in the Professional Paper on which the brief Chapter 9 is largely based.

1 Methods and Results of River Terracing

HUGH MILLER

THIS paper is an essay towards the elaboration of an outline sketch produced by one of the firmest hands that have ever worked in geological landscape, that of Playfair. As he writes:

> When the usual form of a river is considered, with the trunk divided into many branches which rise at a great distance from one another, and these again subdivided into an infinity of smaller ramifications, it becomes strongly impressed on the mind that all these channels have been cut by the waters themselves. The changes which have taken place in the courses of rivers are to be traced in many instances by successive platforms of flat alluvial land, rising one above another, and marking the different levels on which the river has run at different periods of time (Playfair, 1802).

REVIEW OF LITERATURE

Less, perhaps, has subsequently been added to Playfair's original outline than might have been expected. I need only allude to the old cataclysmal views which stood in the way of the Huttonian theory: waves of translation, for instance, and other suppositious floods; sudden emergence of countries from the sea, whereby the waters were thrown violently off; earthquake formation of valleys and lakes, whose waters were drained from level to level by ruptures in the rock below. The land shells in the alluvia proved them to be ordinary deposits, and neither geology nor common sense can exist if rare or imaginary accidents are to be accepted as explanations of what is universal, merely on account of their speed in action.

Many of the early geologists were led insensibly from old coastal terraces to terraces on inland rivers. The transition was found by Darwin (1846) in South America, and by Chambers (1848) in Britain. Both inferred a single origin for coastal and for river terraces. Darwin's coastal terraces ran into river valleys, assuming the slope of these and continuing into terrace-like fringes in the Cordilleras at heights of 7,000 to 9,000 ft. Darwin accounts for the whole series by the progressive lengthening of deltas in response to progressive elevation, the steps of terraces being taken as old sea cliffs. Chambers interpreted Scottish terraces in a similar way, whereas the alternation of freshwater and marine conditions in the lower parts of some valleys is, in actuality, proof to the contrary.

With modifications, the uniformitarians in the controversy about the paleolithic terrace gravels of the Somme and southern England accepted a marine origin of terraces. Thus Prestwich (1864) writes: 'According to any variability in the rate of elevation, to intervals of repose, or to deflections in the flow and velocity of the rivers, so there may exist intermediate terraces or levels, sudden variations in the slopes, and gravels lodged at various levels.'

Lyell, maintaining this same view in 1841 – well before Prestwich – seems never to have qualified it (Lyell, 1871); in Prestwich's day Dana (1863) was working out identical opinions (Fig. 1.1), which he embodied in his textbook. Whitaker (1875), Foster and Topley

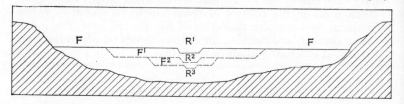

Fig. 1.1 Formation of terraces during successive uplifts (after Dana) R^1, R^2, R^3, successive channels; F, F^1, F^2, successive flood plains.

(1865), Home (1875) and Hull (1878) also explained terraces by means of successive accelerations in excavating power produced by successive elevations of the land.

Hitchcock (1833), in America, seems to have been among the first after Playfair to recognise that rivers will make terraces even if left to themselves. In subsequent writings he distinguishes terraces cut under the control of falling sea-level from those due to a natural tendency of rivers to terrace their banks. In a later work, however (Hitchcock, 1857), he unfortunately represents each terrace as built up separately from the bottom of the valley (Fig. 1.2). He groups the terraces of the Connecticut river into four classes:

1. Lateral, or ordinary, terraces.
2. Delta terraces, intersected deltas at the mouths of tributary streams.
3. Gorge terraces, occurring either above or below gorges, and being intermediate between classes 1 and 2.

4. Glacis terraces, a doubtful group with a slope away from the river in addition to the usual bank on the riverward side.

Hitchcock regarded terraces as important enough to define one of his three divisions of post-Pliocene time. His views have been restated by Upham (1877), who accounts for the highest terraces,

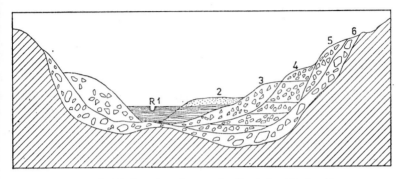

Fig. 1.2 Ideal section of terraced valley (after Hitchcock).
1, Flood plain, with R river. 2 and 3, River terraces. 4, Moraine with kettles. 5, Probable sea terrace. 6, Till.

approximately equal in elevation on the two sides of the valley, by the great flood of the time of glacier decay. Dawson (1881), dealing with British Columbia, appeals to submergence in respect of the high, corresponding, terraces, but relies on river action to explain the lower, non-corresponding, features.

Tylor (1869) has maintained that terraces were produced during a rain period when, for instance, the Yorkshire Aire discharged 125 times as much water as now. Brown (1870) takes a similar view with respect to terraces on the Earn and Keith, while Dana, in his latest writings, discards earlier inferences, ascribing the Connecticut terraces to melt-water floods, although he needs also to postulate changes in gradient. Smaller climatic changes are easier to accommodate. Whitney (1862) appeals to the alternation of drainage and repose. Drew (1873), referring to the Himalayas, calls on variations in the supply of glacier debris. Kjerulf (1871) in Norway distinguishes between terraces in closed situations, which he takes as former lake beaches, and terraces in open situations, where terrace fronts represent old clifflines. Jamieson (1874) offers yet another possibility (Fig. 1.3), that of the lodgement of morainic gravel alongside a melting glacier.

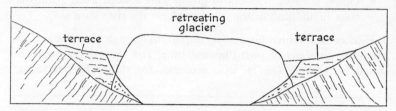

Fig. 1.3 Lodgement of gravelly outwash alongside glacier (after Jamieson).

Summary of possible origins of terraces
1. Ancient sea margins.
2. Ancient river flats deserted on account of elevation.
3. Ancient river flats, episodically incised in response to climatic change.
4. Relics of former monstrous floods.
5. Shorelines of former spasmodically tapped lakes.
6. Uplifted fluvio-marine banks.
7. Morainic debris banked against melting glaciers.
8. Submarine deposits, answering to kames or *åsar*.
9. Features produced by river action during downcutting, when impulses of erosion took effect.

RIVER TERRACING: ITS METHODS

In the excavation of valleys, streams do not generally confine themselves to the narrow limits of a channel along the middle. They extend their operations over some breadth of ground, upon which they wind and shift, erode and deposit, keeping open only a narrow channel, and flooring the rest of the bottom of the valley with alluvial deposits. The processes which produce flat-bottomed valleys are the same as those which produce tiers of level, alluvium-strewn surfaces separated by the risers which convert them into terraces (Fig. 1.4).

Formation of river curves: deflection pools
Winding streams tend to scoop at the elbows of the curves. A straight course does not suit the conditions of a mobile fluid passing over a variable bed. Some inequality in the channel or weakness in the bank is sure to direct the force of the stream against one side, and it then begins to eat out a curve. The apex of the V or the U in the cross-section is forced to one side. The water, heaped up and recoiling,

scarps the bank into a cliff and scoops the bottom into a pool. After being deflected from one bank the stream is projected against the

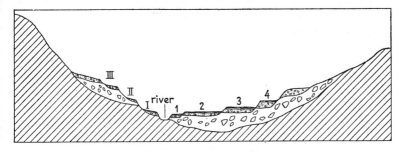

Fig. 1.4 Ideal section of a terraced valley.
1 to 4 and I to III. Terraces of gravel and alluvium, excavated in till and bedrock; highest terrace without alluvium.

other; one curve propagates another. The pools may be termed *deflection pools*.

Travelling of river curves and deflection pools
Since the bends in a stream act as partial dams (Humphreys and Abbott, 1876), and since the water gnaws ceaselessly at the deflecting bank, neither pools nor cliffs are perfectly stationary: they tend on the whole to travel down-valley – in soft materials, at measurable rates. Tylor cites the downstream shift of one cliff on the Ganges as amounting to 150 ft in a single flood.

The travelling of the curves in a river makes it resemble somewhat a rope shaken into horizontal undulations; but in the river the movement is very irregular, on account of variation in the angle of attack on the banks, or inhomogeneity of the bank material. Force and resistance are unlikely to be balanced, even throughout the length of a single curve. Hence the distortion of open curves into closing loops and horseshoe bends. Tortuosity can increase until the two extremities of the horseshoe come together, and the water breaks through, particularly in time of flood, when the elevation of the plane of the water, where it is first checked, can amount to 10 or 12 ft.

Planation
The various methods of river action, including shift, sweep, distortion and cut-off, result in an intermittent planing-off of the strata along the line of the valley. Each river curve, forced horizontally like a curved knife with the blade laid flat, may leave as it travels not only

a plane over which it has passed, but a bank along the base of which it has been drawn. The running water in its circuits may be likened to a scythe in its traverses along the edge of a field; there is a breadth of shorn stubble and a wall of standing corn. Above the stream bed may occur higher planes of river denudation, where truncated strata are now overspread with alluvium.

Relation between the travelling of river curves and formation of gravel
The methods by which surfaces of planation become gravel-covered bear a direct relation to the travelling of the stream curves. All parts of the stream courses which are not kept open by scour are filled with stream sediment; deposition of alluvium keeps pace with the movements of the curves, sedimentation on the inner bank compensating for erosion on the outer (Fig. 1.5).

Deflection pools are usually shallow on the side opposite to the deflecting bank. At the base of the scar on the deeper side can often be seen something of the raw material on which the river is working. The stones in the stream bed on that side are few and large; lying in the thread of the stream, they bear the full force of the current. Towards the far side of the channel the bottom shelves up, the current slackens and the average size of particle becomes smaller, evidently in response to the sorting action of progressively slackening flow.

The following table, taken from Stevenson, summarises the relationship between transporting power and velocity:

Velocity of water		
(in. per sec.)	(m.p.h.)	Effect
3	0·170	Water will just begin to work on fine clay
6	0·340	Water will lift fine sand
8	0·4545	Water will lift sand as coarse as linseed
12	0·6819	Water will sweep along fine gravel
24	1·3638	Water will roll along rounded pebbles 1 in. in diameter
36	2·045	Water will sweep along slippery angular stones the size of an egg

The gravel bank grows by accretion. Coarser materials, with some finer admixed, are added to the lower parts nearest the thread of the current; finer materials fall from the slacker water farther away from this thread; and the first tufts of grass and the first lodgement of sand may appear together, the sand helping the grass to grow and the grass trapping more sand. The Northumberland Coquet provides an example: boundary stones and the O.S. map show that the river has

Fig. 1.5 Section of channel bend, showing accretion deposits.

pushed through 80 yds in eighteen years, during which time the extension of gravel on the one side has kept pace with it. The gravel receives a top-dressing in time of flood.

The gravelly fraction of the alluvium is not deposited in horizontal layers, but is emplaced by accretion on a slope, the finer covering the coarser by a kind of overlap. The sand and loam which, in that ascending order, cover the gravel, occur in overlapping layers according to the height of the floods which formed them.

The calibre of gravel varies according to the force of floodwater; the average thickness of gravel is determined by the average height of flood. Floods of variable range produce an interweaving of coarser deposits with finer; and extraordinary floods leave deposits not at small bends, but at large ones, and possibly only at the largest in the whole valley.

TERRACE FORMATION

As far as possible, Hitchcock's classification will be adhered to in what follows.

Lateral terraces: 1. Amphitheatre terraces at persistent bends, glacis terraces

There are in every stream numbers of what may be regarded as persistent bends, associated with others which may be viewed as shifting. A well-defined persistent bend is one where the stream sweeps round a deep curve, bounded on the convex side by a high, amphitheatre-like bank, with the opposite or alluvial side often benched with terraces rising in succession from the water's edge. Such bends are working into the convex bank.

Accretion on the inner or sheltered side is not uniform, but results in crescent-shaped additions of gravel and alluvium. A secondary channel may in time of flood come into use on the landward side of the deposits, turning the accumulation temporarily into an island. Discarded secondary channels remain as grassy furrows, whereas the selected one is kept open by scour at flood time.

It seems usual for the current flowing through the selected secondary channel to attack its banks in places. Thus in flood after flood the furrow is etched into a step or little terrace. The more its landward margin is eroded, the higher and more terrace-like this side becomes. When the main channel has shifted away and has cut farther down, the secondary channel will finally dry out. Many abortive terraces may be formed for every one that is eventually preserved; but preservation of some means that the sloped side of the amphitheatre becomes benched, with the benches convex towards the hollow of the opposite bank. To this well-marked variety of river terrace, which Hitchcock did not separately distinguish, the name *amphitheatre terrace* may be applied.

Most river terraces in this country are of this kind. Of those mapped by the writer, 70 per cent occupy positions within stream bends, but the proportion actually produced by the means just described is probably less.

Amphitheatre terraces, forming independently in each loop of the river, need not correspond in relative altitude from one loop to another, and in the writer's experience equality of height is exceptional, particularly above flood levels. Again, the unequal and variable height attained by certain lateral terraces, including the variety under specific discussion here, permits a single terrace to vary as greatly in height as if it were several distinct terraces. The number of amphitheatre terraces that are not more or less composite cannot be great.

A variant form of the amphitheatre terrace often assumes the characters of Hitchcock's *glacis terrace*, its surface sloping away from the stream (Fig. 1.6). The secondary channels already described often temporarily divert large parts, or even the whole, of the main stream,

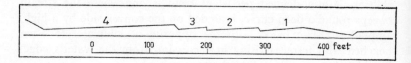

Fig. 1.6 *Glacis terraces on Whickhope Burn, North Tyne.*

so that shifting curves may cut laterally in the oblique plane, which planation, accompanied by deepening, must necessarily follow. Glacis terraces result. The angle of the glacis slopes sometimes amounts to 5° of arc.

Lateral terraces: 2. Junction terraces
At the junction of streams, both currents tend to keep clear of what lies in the angle between them. The gravel spit in this angle may grow into a triangular flat, which periodical destruction at successive levels may convert into terraces. Partial destruction may be effected by meanders on either stream, the loops constructing new flats at lower levels. Here again, heights of terraces differ from one part of the stream to another.

Lateral terraces: 3. Indeterminate varieties. The lateral terrace proper
It is desirable for the terms *amphitheatre terrace* and *junction terrace* to be restricted to forms identifiable as having been produced by the means indicated above. Many amphitheatre terraces, of course, may originally have been formed within once persistent but now lost bends. There is also an ill-defined group of which all that can be said is that they are lateral terraces lacking a determinable relation to persistent bends, formed independently on the opposite sides of streams, and not usually located directly across from one another. The typical member of this group has been lengthened during a long traverse of a single bend (Fig. 1.7).

Fan terraces, or lateral delta terraces
Although Hitchcock applies the term *delta terrace* to a feature which slopes up towards the mouth of a tributary, the term is best restricted to terraces formed at or near intersected deltas. The term *lateral delta terrace* will here be replaced by *fan terrace*.

Deltas are not necessarily mainly submarine or subaqueous. Talus fans, alluvial cones and great river deltas all illustrate a single principle. As Dutton (1880) writes:

> When the stream is progressively building up its bed outside the point of exit, it cannot long occupy one position. The slightest cause will divert it to a new bed, which it builds up in turn, and which in turn becomes unstable and is abandoned. The frequent repetition of these shiftings causes the stream to vibrate radially round the gate as a centre. . . . The formation thus built up is an alluvial cone.

Periodical destruction of alluvial fans by river planation at their margins disturbs the equilibrium. When the tributary encounters the edge of the river bank instead of its own gentle slope, it pours over,

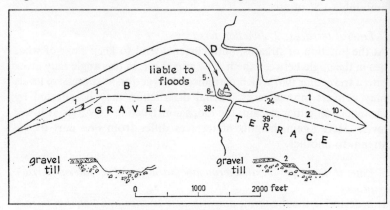

Fig. 1.7 The North Tyne near Wark.
A, Migrating curve of channel. B, Haugh. C, Sandbank. D, Tributary.
1 and 2, Terraces below A. 1′, Lower terrace. 2′, Probable selvage of 2.
Remaining numerals indicate elevation above river (in feet).

cuts a gully back into the heart of the fan, and cannot resume its conical distribution of materials at the earlier level without building up to it again. The old delta is left as a *fan terrace*. Successive interruptions of the fan-forming processes result in a series of fans with their edges clipped into terrace fronts (Fig. 1.8).

The distinctive characteristics of fan terraces are:
1. Radial slopes away from the point of exit.
2. Equivalence of height across the tributary stream.
3. Thinning of the terrace fronts towards either end, in a view from the main valley.

Delta terraces

Like fans, shoreline deltas can undergo intermittent cutting. But the great size and gentle slope of many coastal deltas make them more liable than fan terraces to produce cone-in-cone forms, and to be complicated with other forms of terrace. Upheavals alternating with stillstand, such as are demonstrated by old coastlines, may each obviously result in a kind of *cycle* of river action, containing the following three stages:

1. The stage of the delta terrace, corresponding to upheaval; this is transitional to

2. The stage of lateral terraces, as the stream begins to find lateral play; this corresponds to the earlier part of the stillstand.
3. The stage of a new delta formation within the old, as the stream loses slope and becomes enfeebled; this corresponds to the later part of the stillstand. Subsidence may now recommence the cycle.

Delta terracing in response to repeated upheavals of the coastline tends to extend itself inland: deltas are, in the geometric sense of the

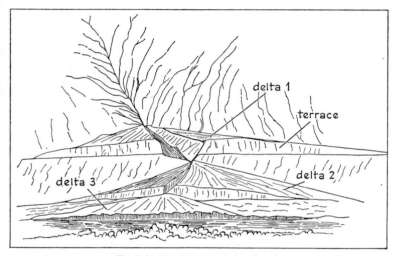

Fig. 1.8 Fan Terraces (after Drew).

term, *produced* in the inland direction. But such extension is probably limited to soft and somewhat uniform materials. A bar of hard rock acts as a regulator in checking spasmodic deepening farther upstream. This circumstance is well illustrated from the Dorback, a tributary of the Findhorn. When in 1838 the Dorback was diverted across the neck of a loop, the course was shortened by about 700 yds. The water in the new cut poured across clay, making a waterfall 15 ft high. In four months the fall had been displaced upstream, through clay till, for a distance of more than 200 yds. In ten months the deepened channel had reached the foot of a rocky ravine, but there the impulse was checked.

Gorge terraces

The writer has not observed examples of the *gorge terrace* of Hitchcock, which is stated to occur either above or below gorges, and to be intermediate between the lateral and the delta (fan) terrace. It is higher than the lateral terrace. In issuing from the gorge it seems to slope away to the normal elevation, somewhat like a fan terrace. But there seems no real need to distinguish, as Hitchcock does, between lateral terraces and gorge terraces. In narrowed positions streams rise higher, and their terraces, if any, must also rise. Thus the Findhorn at Freeburn, where it is 600 ft wide, rose 17 ft during the Morayshire floods; at Randolph's Bridge, where the width ranges from 8 ft to 70 or 80 ft according to stage, it rose 50 ft; and at Craig of Coulternose, downstream of Sluie, a width of 185 ft produced a rise of only 15 ft.

VALLEYS VIEWED IN THEIR RELATION TO RIVER TERRACING

The course of a great river is distinguished by Mr Archibald Geikie in three regions: the *mountain track*, where the young stream, continually swelled by lateral torrents, dashes down mountainsides and through ravines; the *valley track*, where its course is more leisurely, and its rocky parts exist chiefly as gorges between wider and more alluvial reaches of valley; and the *plain track*, where the river winds out on to alluvial plains, largely of its own forming, and often deposits more than it erodes.

In the mountain track streams generally occupy themselves in deepening their rocky V-shaped channels, and in slowly planing them – structure permitting – into an increasingly flat-bottomed U. The only chance of terrace formation in the plain track lies in upheaval of a coast having sufficient slope to communicate activity to the stream. It is in the valley track that terrace formation can best be studied in active progress. The gorges of hard rock by which most valleys are interrupted divide them into compartments where planation of softer material is constantly in progress.

Streams in glaciated areas can run entirely over glacial deposits, entirely over bedrock, or partly over bedrock and partly over glacial deposits, and it is possible for a particular reach of channel to have one bank cut in bedrock and the other in till. Compartmented valleys in glaciated areas in part reflect pre-glacial relief, in part relate to glacial deposition. On balance, glaciation seems to have multiplied the number of gorges. The significance of glacial events for river terracing is immense:

1. Since gorges give rivers leisure for planation in the soft glacial

deposits upstream, rivers can form strips of alluvium too wide to be kept level. In some cases, however, the rock in the gorges is so resistant that downcutting is negligible, so that the movements of the rivers upstream are almost entirely lateral.

2. Modern rivers, striking rock at variable depths, become rock-bound when they do strike it. They are then prevented from further pursuing their work of terrace building and terrace destruction.

These circumstances are illustrated in Fig. 1.9. The later of the terraces there shown are grouped in two compartments, separated

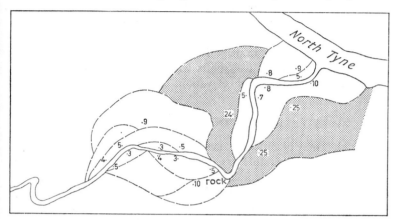

Fig. 1.9 Terraces at mouth of Nunwick Burn, North Tyne; numerals indicate elevations in feet.

by a few yards of rock where terrace building has been arrested. There was, however, a time when the rock lay concealed by till, and when the stream was free to move over the surface of upper terraces, one of which is marked by shading in the diagram.

3. The highly variable character of the valley bottom has encouraged the streams to work into tortuous courses full of persistent bends and amphitheatre terraces.

4. The confining action of the bedrock strengthens the tendency of valleys to narrow with increasing depth. Too little has hitherto been made of this tendency. It has been assumed that, because rivers now occupy narrowed valleys flanked by broad terraces, they have shrunk. But immediately after glaciation the rivers commenced to work on shallow wide-bottomed valleys, where they were able to plane far and wide, travelling from breadth to breadth to an extent never now equalled. With banks now greatly heightened and increasingly rock-bound, they are certain to concentrate their channels

as they excavate them, unless the rate of planation is out of all proportion to the rate of deepening.

5. The division of valleys into gorges and basins has tended to regulate the effects of spasmodic elevation at the coast. It seems impossible that fall of sea-level can have been followed by downcutting all along the rivers. Pulsations and fluctuations doubtless occur. There are cycles in the meteorological causes that produce floods. There may be alternations of harder and softer rock in the barriers. And when a waterfall has worked back to the edge of an alluvial flat, the circumstance doubtless produces some leisurely acceleration in its excavation. But it is hard to escape the conclusion that the effects of these various pulsations are lost, like the effects of pulsations of all kinds, in response to travel through a succession of different media. Successive uplifts will, in the interior, blend their effects into those of equable elevation.

RIVER TERRACING IN MINIATURE: CONCLUSION

Although it may be unwise to generalise from observations in a small part of the world, it is still fair to distinguish between isolated observations and essential principles. It was here that the strength of the Huttonian theory lay. The action of running water is in all times and places guided by a single set of principles.

There is much to be learned from the miniature stream terraces that may be seen on a sandy seashore when the tide has ebbed, or on sandbanks at riversides, and on spreads of deposit sent from the mouths of drains during a thunder-shower. The writer has recently observed an instance of miniature terracing. It occurred on a small runnel which, issuing from a spring, flowed after 10 yds over a smooth beach of fine sand at the edge of a river. Like larger streams, this too had its curves and deflection banks, at the base of one of which it was scooping away sand grains. Opposite the little cliffs were terraced slopes, left on the one side because the stream had pushed to the other: these were amphitheatre terraces in process of formation.

In this case the supply of water was constant. The essential effects of terracing had been produced by the streamlet:

1. Without any fall from terrace to terrace induced by repeated falls in the level of the river.
2. Without the valley's being filled by the stream up to the rim or base of the highest terrace, and hence without a spasmodic fall from one terrace level to the next.

Methods and Results of River Terracing

3. Without the valley's having been so deeply flooded during excavation that the terraces were formed as indented flood margins.
4. Without the stream's being sometimes glutted with sediment too abundant to remove, or alternating between times of glut and times of intersection of surplus deposits.
5. On the contrary, simply by planation at different levels and to varying breadths, with a tendency for the stream to narrow the field of its operation as it deepened its course.

It follows that the existence of terraces is compatible with absolute constancy of discharge. The mere breadth of terraced valleys is no criterion of the volume of water that has occupied them, any more than the $1\frac{1}{2}$-mile depth of the Grand Canyon is any proof that it once held an equal depth of river. The evidence of pluvial and torrential periods must be sought in other directions.

In the best-known valleys the highest line of river gravels, 60 or 70 ft higher than the present bed, is not a terrace, but a shelving, shore-like slope. It would seem that here the first flowing of the post-glacial rivers was not accompanied by terracing. Their shores may have been heavily armoured with ice; or, more probably, the rivers may have been frozen almost solid, being so charged with broken ice that the water was forced to break over the top, sweeping the shores with levelling violence.

In examining evidence for ancient floods, we have a right to expect deposits proportioned to the magnitude of these, and also to expect the scooping of vast scars at the larger turns of valleys and the accreting of immense gravel banks, with huge blocks in them, on the opposite side. The effects of the unusual floods of the present time reveal nothing of what would happen if such floods were the rule.

The opinion seems to be gaining ground among the younger geologists of North America that there, as in Britain, there was but one short torrential period which swept the valley banks and left bare slopes for subsequent river action to terrace. And the first question to be answered by the student of terraces is: To what extent has subsequent action been that solely of rivers, resulting in terraces *not* opposite? It is not permissible to appeal to elevation of the coast, climatic change or periodicity of any kind, without first proving that the terraces range in opposite pairs. If it should ever be proved to the satisfaction of all geologists that the glacial period in this country closed other than with submergence, it will then be admitted that the renewal of free drainage on country with fresh gradients implies everything necessary to terracing as we now observe it.

In this country, as in other northern lands, the *terrace period* is emphatically a post-glacial period. But like all other features of surface configuration, terraces are only temporary. At first steep scars, then assuming the slope of fallen material, they are, as time goes on, being washed and wasted into mere gravel slopes, like the obscure gravel platforms of the 'high-level' gravels in the south of England. By the time they have been effaced, the gradients of the rivers will perhaps have become so low, and their powers so enfeebled, that we may not have other terraces to replace them, until the country enters upon a new phase of geological activity, and corresponding activity is induced in the streams.

REFERENCES

BROWN, T. (1870) 'On the old river terraces of the Earn and Teith, viewed in connection with certain proofs on the antiquity of man', *Trans. Roy. Soc. Edinburgh*, XXVI 149–76.

CHAMBERS, R. (1848) *Ancient Sea Margins as Memorials of Changes in the Relative Level of Sea and Land* (Edinburgh, W. & R. Chambers; London, W. S. Orr Ltd).

DANA, J. D. (1863) *Manual of Geology* (Philadelphia, Theodore Bliss).

DARWIN, C. (1846) *Geological Observations on South America* (London, Smith, Elder & Co).

DAWSON, G. A. (1881) 'Additional observations on the superficial geology of British Columbia and adjacent regions', *Quart. J. Geol. Soc. Lond.*, XXXVII 272–85.

DREW, F. (1873) 'Alluvial and lacustrine deposits and glacial records of the Upper Indus basin: Pt. I. Alluvial deposits', *Quart. J. Geol. Soc. Lond.*, XXIX 441–71.

DUTTON, C. E. (1880) *Report on the Geology of the High Plateau of Utah* (Washington, D.C., U.S. Dept. of the Interior) 307 pp.

FOSTER, C. LE N., and TOPLEY, W. (1865) 'On the superficial deposits of the valley of the Medway, with remarks on the denudation of the Weald', *Quart. J. Geol. Soc. Lond.*, XXI 433–74.

HITCHCOCK, E. (1833) *Report on the Geology, Mineralogy, Botany and Zoology of Massachusetts* (Amherst, J. S. and C. Adams) 700 pp.

—— (1857) *Illustrations of Surface Geology* (Washington, D.C., Smithsonian Institution) 155 pp.

HOME, D. M. (1875) 'Notice of some high-water marks on the banks of the river Tweed and some of its tributaries; and also of drift deposits in the valley of the Tweed', *Trans. Roy. Soc. Edinburgh*, XXVII 513–62.

HULL, E. (1878) *Physical Geology and Geography of Ireland* (London, Edward Stanford).

HUMPHREYS, A. A., and ABBOTT, H. L. (1876) *Report upon the Physics and Hydraulics of the Mississippi River* (Washington, D.C., U.S. Army Corps of Engineers) Prof. Paper 13, 691 pp.

JAMIESON, T. F. (1874) 'On the last stage of the glacial period in north Britain', *Quart. J. Geol. Soc. Lond.*, XXX 317–41.

KJERULF, T. (1871) 'Om terrasserne i Norge og deres betydning for tidsregningen tilbake til istiden', *Geol. Mag.*, VIII 74–6 (see also *Skand. Naturforsk. Forhandl.*, X (1869) 631–75; *Deutsch. Geol. Gesell. Zeitschr.*, XII (1870) 1–14; and *Halle Zeitschr. Gesamt. Naturwiss.*, II (1870) 496–8).

LYELL, C. (1871) *Student's Elements of Geology* (London, John Murray).

PLAYFAIR, J. (1802) *Illustrations of the Huttonian Theory of the Earth* (Edinburgh, William Creech) 528 pp.

PRESTWICH, J. (1864) 'On the Loess of the Valleys of the South of England, and of the Somme and the Seine', *Phil. Trans. Roy. Soc.*, CLIV 247–309.

TYLOR, A. (1869) 'On Quaternary gravels', *Quart. J. Geol. Soc. Lond.*, XXV 57–100.

UPHAM, W. (1877) 'The northern part of the Connecticut valley in the Champlain and terrace periods', *Amer. J. Sci.*, 3rd ser., XIV 459–70.

WHITAKER, W. (1875) *Guide to the geology of London and the Neighbourhood*, Mem. Geol. Survey of England and Wales (London, H.M.S.O.).

WHITNEY, J. D. (1862) 'On the Upper Mississippi Lead Region', in J. Hall and J. D. Whitney, *Report on the Geological Survey of the State of Wisconsin*, I (Albany, Geol. Survey of Wisconsin) 455 pp.

2 River Terraces in New England
W. M. DAVIS

I. GENERAL STATEMENT

Theories of river terraces

IN the study of the terraces carved in the washed drift of New England valleys, more attention has hitherto been given to cross-section than to plan. The cross-section is usually represented as in Fig. 2.1, where the cross-valley distance between terrace scarps at low levels is less than the corresponding distance at higher levels. This circumstance has frequently been taken to indicate that the volume of present-day streams is less than the volume of the streams which carved the high-level terraces. It will, however, be shown that the characteristic terraced section cut wholly in drift contains few steps (Fig. 2.2); if steps are numerous, then many terraces are typically defended at the base, over short distances (10 to 50 ft), by rock ledges (Figs. 2.3, 2.4).

When the terrace pattern is considered in plan as well as in cross-section, it appears that the New England terraces may be accounted for by (1) degradation effected in a previously aggraded valley, without change of stream volume; and (2) the control of lateral swinging by rock ledges, as suggested by Miller (1883).

II. PRELIMINARY INQUIRY

Various kinds of terraces

The New England terraces are known as river terraces, drift terraces or alluvial terraces. They have nothing genetically in common with terraces of sea or lake shore. They bear little resemblance to structural benches such as those of the Colorado Canyon. They have little similarity to the rock terraces with covers of silt and gravel, formed when a graded river, revived by uplift, cuts a new valley in its former floor, as along the Rhine gorge (Fig. 2.5).

The New England terraces have flat and nearly level upper surfaces, limited backwards by rising ground and forwards by falling ground.

They are evidently the river-carved remnants of bodies of surficial deposits contained in rock-floored valleys of still earlier origin. Their upper surfaces slope down-valley with the slope of the streams which

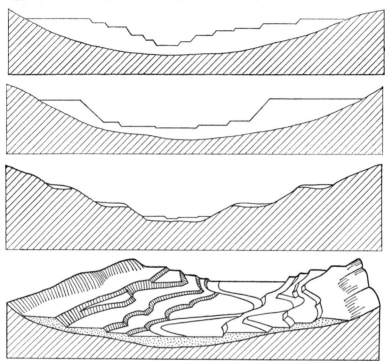

Figs. 2.1–2.4
Sections of terraces carved in the washed drift of New England.

have cut their scarped fronts. They consist of unconsolidated, stratified drift; if a ledge of bedrock appears, it is manifestly an accidental element, although, as will be shown, capable of controlling the pattern of terrace fronts. These drift terraces are usually less regular in pattern than are terraces of other kinds. A single drift terrace, unless it be the highest one of a series, is seldom traceable for many

Fig. 2.5   Sections of terraces carved in the Rhine valley.

miles along the valley, and it may be merely a few hundred yards in length. Terraces of other kinds are usually much more persistent.

Furthermore these drift terraces differ from other terraces in the place that they occupy in the geographical cycle. They are not the products of normal erosion during an undisturbed stillstand, but the consequence of a short-term departure from normal progress. The valleys which contain the New England terraces imply the previous attainment of maturity. They were modified during the glacial period, which closed with the accumulation of abundant drift and with certain changes of level whereby erosion was promoted. Even when the drift has been cleared, the normal cycle of erosion will not have advanced far beyond its pre-glacial phase.

Terrace patterns

The plain or floor of a drift terrace frequently varies rapidly in width, usually terminating in points at the two ends (Fig. 2.4). Its borders form curves of tolerably uniform radius, concave to the stream and frequently uniting in cusps. When several cusps are grouped in a strong salient they may be called a *terrace spur*. The highest plain of a flight of terraces backs against the older valley side, while each lower terrace backs against the scarp of the next higher. The back border of a terrace is frequently followed by a marshy channel from which the terracing stream has been withdrawn; thus terrace plains may slope gently away from the axis of the valley. Terraces of this kind are called by Hitchcock (1857) *glacis terraces*.

The scarp of a terrace presents a succession of curved re-entrants separated by salients, carved by the successive encroachments of a stream which has swung laterally at least as many times as there are terraces. Although the plain and the descending scarp at its front are usually taken together as bounding a terrace, these two surfaces are not genetically connected in river terraces; it is the ascending scarp at the back which should be associated with the plain beneath and in front. The line between the plain and the ascending slope is the most significant of all terrace lines. The line at the front is determined merely by the slipping of the drift.

Terrace scarps are steepest where the stream has most recently swung against their base. The youngest and steepest scarps are at the bottom of the flight, except where a chance lateral swing undercuts even the highest terrace plain; then the whole descent from highest to lowest level may be fresh-cut. In older terraces the scarps weather to a gentle slope, and the edges are rounded off. A convex slope of erosion is thus formed above and a concave slope of deposition below. Gulches are often worn in terrace fronts by wet-weather streams, and fans are spread on the terrace plain below, while the abandoned channels at the back borders guide surface drainage (Fig. 2.6).

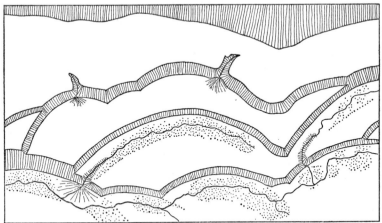

Fig. 2.6 *Fans, terraces and abandoned channels.*

Terraces carved by streams of diminishing volume

The primitive explanation of terraces is that the whole space between the upper terrace scarps represents the channel of a huge river, and that successive diminutions of volume are indicated by the descending decrease of breadth between the lower terraces. Although this view has never gained general acceptance in the strict form, it has nevertheless been generally supposed that present-day rivers are much smaller than were their precursors at the time when terracing began.

Now the best indication of the volume of the stream which carved a particular terrace is the curvature of a terrace scarp. If the radius and arc of curvature of low-level terraces are similar to these elements in the meanders of the actual stream, and significantly smaller than in the high-level scarps, while curves at intermediate levels show intermediate values, then a diminution of stream volume may be inferred. If the magnitude of radius and arc is similar throughout, no change of volume is indicated.

On the other hand, if a stream were charged with abundant coarse load in the last stages of aggradation – as seems to have occurred in New England – its slope must have been relatively strong; and a graded river with a heavy load on a strong slope does not develop curves of as small radius as it would with the same volume but a finer load on a gentler slope. Hence a large radius of curvature in the uppermost terraces is not alone sufficient to indicate larger volume; a large arc of curvature is also needed. It is for this reason that some of the uppermost terraces on the Connecticut and Westfield rivers, where radius of scarp is greater than at low levels, do not provide adequate evidence for change of stream volume.

At the same time, what is known of the later stages of the glacial period makes it very probable that the New England streams had in fact greater volume then than now. Water was turned from the basins of north-directed rivers into valleys running southwards. When the northern basins were still ice-covered, subglacial streams may have been forced to cross the divides toward the south. After the ice had partially withdrawn, ice-dammed lakes may have discharged across low points into the southern valleys. Another possible cause of increased volume is the importation into southern basins of a considerable snowfall received on the ice sheet to the north. Again, it is possible that precipitation was greater during the later stages of the glacial period than at present. Finally, rapid melting of the retreating ice may have swollen the southern rivers. It is entirely possible that these various causes may have increased stream volume while the New England valleys were being aggraded; but it does not follow that the increase persisted into the period of terracing.

Except where direct evidence for it is provided by curvature and arc of high-level terrace scarps, a formerly greater volume of the terracing streams must be taken merely as possible, not as demonstrated. Similarly bulk and coarseness of the terrace deposits should not too readily be accepted as evidence of former great volume. Bulk is a function of time as well as of rate of action, while texture is a function of slope as well as of volume and velocity. Until bulk and texture are known to be insufficiently explained by the time available, they do not necessarily indicate increased volume.

And even if a general decrease of volume has occurred during the period of terracing, it has nevertheless not controlled terrace development. If it had, stepping terraces should be much more abundant than they are. In actuality, it is rare to find a long flight of stepping terraces on both sides of a valley, or a particular flight continuing for any long distance. It is not uncommon to find at least one side of the valley enclosed by a single scarp. If terracing had been due to a general decrease in volume, broad flood plains between the high scarps of a single terrace on each side of the valley should be much more rare than they are, and when the whole descent is made in a single scarp on one side, then stepping terraces should be well developed on the opposite side. No such arrangement, however, can be said to prevail. Decrease of stream volume must therefore be at most a subordinate cause of terracing, if not indeed as a rule negligible (cf. Adams, 1846, pp. 145–6).

Terraces carved by streams of increasing slope
When the basin of an aggrading river system is slightly tilted, streams whose slopes are decreasing can be expected to aggrade their valleys more rapidly than before, while streams with increasing slopes will change from aggrading to degrading. New England is well known to have undergone differential elevation in post-glacial times: the post-glacial clays of Lake Champlain and southern Maine were deposited when the sea stood 300 ft or more above its present level, but no post-glacial change of level of such an amount is known along the southern New England coast. South-flowing rivers have therefore been accelerated, north-flowing rivers retarded. But while it thus becomes very probable that the erosion of valley drift was determined by unequal post-glacial uplift, it does not follow that individual terraces are at all closely related to the movement.

The northern uplift may have been a single rapid movement, causing the streams to deepen their valleys quickly for a time, and to defer lateral swing of the kind essential to terracing until they had developed new grades of gentle declivity. In this case only a single high-level terrace and no intermediate terraces would be formed, while terraces at low levels would be few.

Repeated uplifts separated by pauses should cause the rivers to undertake lateral swing at as many levels as there were pauses. The flood plain formed in a given valley during each pause would be relatively persistent along the valley. But in order to account for the preservation of parts of the higher terraces, it would be necessary to postulate that movements of uplift succeeded one another at shorter and shorter intervals, so that the later-carved flood plains could be narrower than the earlier ones. The chief objection to this possibility in the present context is that it requires correlated levels on the two sides of a valley, whereas such correlation is by no means characteristic of the New England area; discordance of level is usual.

If uplift were slow, and incapable of greatly accelerating the south-flowing rivers, the larger of these might continue to swing while all the time degrading the valley floor. Terraces could then be cut at many different levels on opposite sides of the valley. This supposition seems most appropriate to New England. The fact that small streams frequently show only faint terraces or none at all seems to mean that the uplift was too fast for the little streams to keep pace. Even the largest rivers have not been able to maintain graded channels in the occasional rock ledges which they have met, while till, intermediate in resistance between stratified drift and bedrock, is capable of preventing the opening of a valley floor, even though a stream may be graded where it flows through the till body.

While slow uplift is thus consistent with the production of many terraces, it is not consistent with their preservation. It fails to explain the downward diminution of the interscarp space. Indeed, the present rivers might tend to develop broader flood plains by strong lateral swinging at the faint grades now assumed than at the stronger grades of earlier times of more active uplift and heavier load. This would mean the undercutting of the earlier terraces, and the formation of a single terrace front, as in Fig. 2.2. In order to explain the preservation of high-level and intermediate terraces it is necessary to introduce some additional factor.

Terraces carved by streams of diminishing load

A graded river may be caused to degrade as well by diminishing its load as by increasing its slope, volume remaining constant. A diminution of load since glacial retreat is highly probable. The streams must accordingly have set to work to degrade the valleys that they had previously been aggrading, even if no change of slope took place. The process, if working alone, must have been very gradual, but it still does not explain stepping terraces. A reduction of the rate of downcutting would again cause the streams to destroy all earlier terraces by broadening their flood plains to the maximum.

Preservation of terraces by rock ledges

Miller (1883) was the first to make explicit the idea that a swinging river will, in cutting down, encounter bedrock from time to time, and that the deeper it cuts the narrower will be the width of fill where free swinging is possible. Even Miller, however, does not accord rock ledges the importance they deserve. For New England, Adams (1846) and Hitchcock (1857) have both drawn attention to the association of particular flights of terraces with occurrences of bedrock, without, however, reaching out to a general statement. Emerson (1898), although fully recognising the defensive function of rock ledges, is concerned chiefly with trenches cut wholly into bedrock, rather than with isolated protruberances. A detailed scrutiny of the whole matter is required; it is attempted below.

Origin of terraces in New England: summary

1. Diminution of stream volume may have taken place during the terracing of the New England valleys, but has not been essential to the production of the terraces observed.

2. The terracing rivers have slowly degraded their aggraded valleys while swinging from side to side, in combined response to a northern uplift and a gradual decrease of load, and in spite of a probable decrease of volume.

3. The chance discovery of rock ledges by the swinging river is the chief cause of the systematic diminution of interscarp space and of the preservation of terraces.

III. THE THEORY OF RIVER TERRACES

The immediately preceding summary gives a guide to the proposed approach of the remaining parts of this paper. The deductive character of the immediately following paragraphs is more apparent than real. Many features of river work presented here as deductions were discovered by observation. It is chiefly for the sake of a continuous presentation of the theory of terraces that a deductive treatment is chosen.

Behaviour of a wandering river

The diagrams introduced below represent several successive stages in the process of slow degradation by a wandering river. The postulates of behaviour on which they are based are that (1) the degrading stream maintains an essentially graded condition; (2) lateral swinging of the channel is much faster (e.g. a hundredfold) than the lowering of the valley floor; (3) the breadth over which a free river tends to swing laterally is greater than the breadth of the meander belt; (4) an individual meander tends to enlarge its radius and to work its way down the valley until it undergoes cut-off.

1. *Maintenance of grade.* If the ratio of load to carrying power (volume and slope) changes but very slowly, a river may remain in an essentially graded condition all through the process of aggradation or of degradation, and through the change from one to the other. It is true that the graded condition depends on a balance between load

and carrying power, and it would at first sight appear that any change in either would throw the river out of grade. But if the change is only by a quantity of the second order, adjustment will follow so immediately that no failure of adjustment will be noticeable.

2. *Rates of swing and of cutting.* A graded stream may continue to work at the lateral wear or building of banks, however slowly it aggrades or degrades its valley floor.

3. *Breadth of swing.* Lateral oscillation carries many rivers from side to side on a flood plain much wider than the meander belt; on the Mississippi the meander belt is 6 to 8 miles wide where the flood plain is 20 to 60 miles wide.

4. *Enlargement of meanders, shift, and cut-off.* On a meandering stream in a broad flood plain, the thread of fastest current runs near the outer banks of curves, where it determines the line of greatest depth. On passing from a given curve this thread is necessarily delivered to the next downstream tangent on the down-valley side of the channel; only after flowing for a significant distance will it attain mid-channel (Fig. 2.7). Bank erosion occurs along the length of contact with the fastest current, while deposition takes place along the bank from which this current is withdrawn. The stream tends to wear away the bank on the outer side of curves, and also on the down-valley side of short tangents. The curves thus increase in radius and arc, and the meander belt widens; at the same time each individual meander tends to move slowly down-valley. The flood plain is scoured away along the concave banks and along the up-valley side of each lobe, while scrolls of new flood plain are added at the ends of lobes and on their down-valley sides.

Similar down-valley shifting is seen in the enclosed meanders of many rivers, typified in the north branch of the Susquehanna. The upland spurs that enter the curves of this river, subjected on their up-valley sides to persistent sweeping, have strong bluffs. But in this rock-walled valley the down-valley shifting does not seem to have been more than fifteen or twenty times greater than the degrading of the river channel in the latest period of trenching, whereas in drift-filled valleys the first of these changes exceeds the second in a much higher degree.

A natural limit is set to the dimensions of a growing meander curve on a flood plain by the formation of short-cuts. Channels abandoned by cut-off can be expected, on the average, to show a larger radius of curvature than the curves of the existing river; this expectation is to some extent confirmed by maps of the Mississippi.

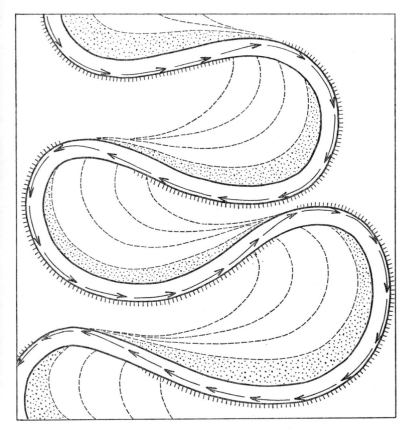

Fig. 2.7 The thread of the fastest current in a meandering channel.

Terminology of wandering rivers
The terms already introduced may now be summarised and somewhat extended:

1. *Meander belt:* the space enclosed between tangents drawn outside the meanders.
2. *Sweeping:* the progressive down-valley movement of meanders.
3. *Swinging:* lateral movement of the meander belt from one side of the valley floor to the other; however, it is not always possible to distinguish between true swinging on the one hand, and the more local shifting of an irregular sweeping meander on the other.

4. *Wandering:* the compound movement of sweeping meanders in a swinging meander belt.
5. *Belt of wandering:* the whole breadth of valley floor that may be worn down by the stream; this corresponds to many New England flood plains.

Ideal terrace pattern: early stage

If the wandering stream is slowly degrading its valley floor, each meander will sweep past a given point at a slightly lower level than that of its predecessor; and each time the meander belt swings across the valley from one side to the other and back again, it will return at a distinctly lower level than the level at which it left. It is the remnants of flood plains so formed that constitute the terrace plains under discussion.

Soon after the stage of degradation has been definitely established, and the meandering stream begins to swing across the valley, a condition such as that represented in Fig. 2.8 may be reached. Terrace *B* is

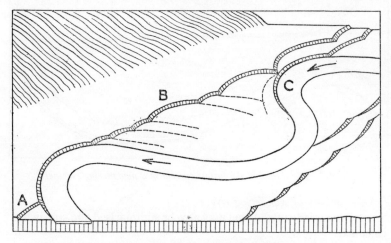

Fig. 2.8   A meandering stream beginning to swing across a river valley.

of greater height than *A*, because *A* has been undercut and consumed in the production of *B*. Terrace *C* is as yet independent of *B*, and therefore rises only by the few inches or feet which correspond to the difference in the level of two successive sweeping meanders. The curves and cusps of terrace *B* result from a vacillation of the meander during its down-valley progress; some cusps are sharp, others join in an almost straight front.

Scarps of the kind shown along the front of terrace *C* in Fig. 2.8 may be called one-sweep scarps, and their cusps may be called one-sweep cusps.

Heights of one-sweep scarps are typically small and may be scarcely perceptible. But if, after carving a series of such scarps, the river now swings to the opposite side of the valley and goes to work there, many meanders may sweep down-valley before it swings back again. When it does so it will be working at a lower level than before, so that, as it once more undercuts the high plain, it will produce a distinct terrace with a scarp of 10 or 20 ft. Terraces of distinctly different levels may therefore usually be taken to represent different swings of the meander belt.

Ideal terrace pattern: middle stage
If the sweeping and swinging of the river continue, a terrace pattern of some complication may result. Few and small remnants of the higher terrace plains are to be expected, but larger and more numerous remnants of the lower plains may be preserved (Fig. 2.9). The

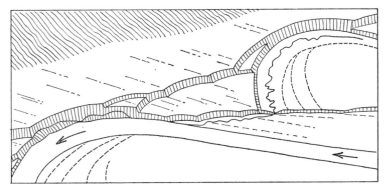

Fig. 2.9   Terrace remnants.

greater the number of swings, the smaller and rarer will be the remnants of the higher terrace plains, unless some special control is present to preserve them.

A special interest attaches to the form and arrangement of the cusps produced by the chance intersection of terrace fronts. In two-sweep or two-swing cusps, two kinds of pattern may be produced (Figs. 2.10(*a*) and (*b*)). These may be called two-swing or two-sweep cusps, with an upstream or downstream *Y*-stem, as the case may be.

A series of one-sweep cusps may occur with some regularity along a valley side, but two-sweep and two-swing cusps cannot be expected to show so definite an arrangement, unless under the control of something more systematic than the action of the wandering stream. It is evidently desirable to analyse the special configurations that may

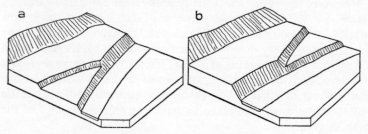

Fig. 2.10 Two-swing cusps.

be due to river action alone, in order to detect the patterns that must be referred to some other cause.

A later-made one-sweep cusp may occasionally chance to stand in front of an earlier one-sweep cusp (Fig. 2.11 (*a*)), but it is out of the question that four cusps should gain a systematic position of this kind (Fig. 2.11 (*b*)) without some special control. Again, while two-swing cusps are common, three-swing cusps (Fig. 2.12 (*a*)) are rare, for they involve the intersection of three unrelated lines; and four-swing cusps (Fig. 2.12 (*b*)) are practically impossible.

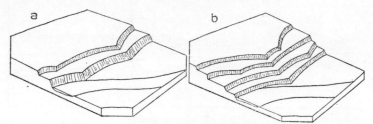

Fig. 2.11 One-sweep cusps.

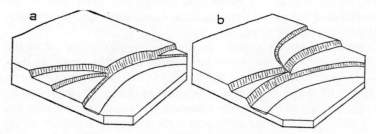

Fig. 2.12 Three-swing and four-swing cusps.

Ideal terrace pattern: late stage

When degradation weakens and ceases, the stream will repeatedly swing to and fro on about the same plane. Even the basal terraces may then be almost completely swept away (Fig. 2.13), and the whole

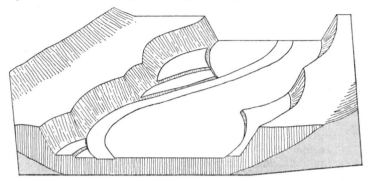

Fig. 2.13   Removal of terraces in a valley.

descent from the high-level terrace to the flood plain united in a single strong escarpment. The required conditions are: the attainment of nearly fixed values of volume and load, such as might be reached when a glacial climate had given way to a milder; the cessation of any slow uplift by which degradation had been initiated or aided; and superposition of the stream on strong rock where corrasion is very slow. A stream would then almost cease to degrade its channel, and would devote practically all its energy to lateral cutting. It would wander back and forth across its valley floor until it came against a lateral terrace, which itself would be worn back to the limit of wandering. Sooner or later the stream would consume all high-level and intermediate terraces, pushing back their united scarps into a single scarp.

Defended terrace cusps: early stage

It has thus far been tacitly postulated that no buried ledges should be discovered by the wandering river. But now suppose a series of terraces to be developed, in which ledges shall here and there be discovered as the river degrades its valley floor to greater and greater depths. It is evident that the number of such ledges may vary greatly; but in all cases the slope of a ledge face will seldom be as steep as the average slope of a terrace front, which may be as much as 30° in freshly cut scarps.

As before, the river wanders about freely so long as it is working on unconsolidated sands and clays; but when a ledge is encountered in the river bank, as at the left forward edge of Fig. 2.14 (*a*), the rock is practically indestructible. The stream will in a comparatively short

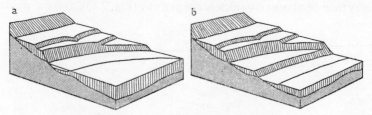

Fig. 2.14   Incomplete terrace destruction.

time swing away; the ledge thus comes to determine a cusp in the terrace front. A salient of this kind may be called a defended cusp, as distinct from the accidental or free cusps described above; the terrace behind it cannot be destroyed by the stream.

The ledge does not determine the depth to which the river may work; the rock is exposed only in the river bank and enters but a little distance into the channel. The slope of the river and the depth to which it has cut are determined by the maintenance of an essentially graded channel with respect to some controlling base-level farther downstream.

When the withdrawing stream swings back again at a lower level (Fig. 2.14 (*b*)), it cannot often undercut and destroy all of the terrace on whose back border the first ledge rises, because (as noted) the slope of the ledge is seldom as steep as that of the terrace scarp. A second encounter with the ledge will usually be made before the swinging stream has entirely consumed the terrace of the previous swing. Every return of the swinging river against a sloping reef of rocks will thus be recorded by a little strip of terrace, and a flight of stepping terraces will necessarily be produced wherever a large group of ledges slopes under the valley drift into the belt of river action. The effect of a ledge will be prolonged by a trailing terrace, as it may be called, stretching far along the valley side.

Slipping meanders and blunt cusps
It may be inferred from the forms of terraces that a stream has two methods of making its passage past a ledge.

Let it be supposed that the river in Fig. 2.15 (*a*) is making its fourth swing against the western side of the valley, and that a buried ledge lies close inside the group of free cusps in the middle of the diagram. The ledge is discovered in Fig. 2.15 (*b*); it lies somewhat below the apex of a downsweeping meander. Assuming that the meander curve is to remain practically unchanged, it can pass the ledge only by withdrawing towards the axis of the valley; it may thus, as it were, slip by the obstacle. Fig. 2.16 (*a*) represents the stream as having just slipped past the ledge, Fig. 2.16 (*b*) as having swept somewhat farther down the valley. All records of the first and third swings of the river

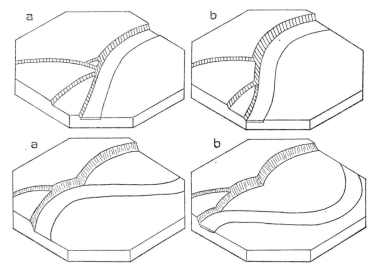

Figs. 2.15–2.16   Terrace development around an obstacle.

are now destroyed at this site. The terrace front shows a high, defended, one-sweep cusp, a free two-sweep cusp with an upstream Y-stem, and a free one-sweep cusp.

The ledge at the base of a defended cusp may come to be more or less concealed by the sands that are washed down from the weathering scarp. Some apparently free cusps may really be defended cusps, with their defending ledge ambushed beneath a thin cover of soil.

Compressed meanders and sharp cusps

This combination represents the second method of passage. Let it be supposed that the river in Fig. 2.15 (*b*) is unable to slip past the ledge. The front of the curve is held fast; the apex bends outward, cutting a curved re-entrant in the terrace front next upstream (Fig. 2.17 (*a*)), while the meanders farther upstream continue their advance. The meander next to the ledge is therefore compressed to a strong curvature (Fig. 2.17 (*b*)). The defended cusp is now sharpened. The compressed meander cannot slip by the ledge; there is no escape save by a short-cut (Fig. 2.17 (*c*)). A sharp cusp

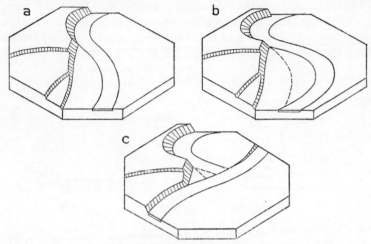

Fig. 2.17 Compression of meander curvature.

is produced, the great concave scarp adjacent to it having an abandoned channel at its base.

Terrace fronts near defended cusps
The difference of behaviour between slipping and compressed meanders depends on the position of the obstructing ledge. If the ledge lies near the apex of the meander, the stream may slip past, but if the ledge is encountered near the point of river inflection between two meanders, compression of the meander upstream from the obstacle is likely to result.

Defended cusps: later stage
After a ledge has once been discovered by the swinging river, the forward reach of its underslope will probably present an obstacle to the stream during every swing towards the valley side. Fig. 2.18 (*a*) shows the blunt cusp of a slipping meander thus determined. Another swing out and back having been accomplished, Fig. 2.18 (*b*) shows the work of a compressed meander, which for a time was held up-valley from a third exposure of the long-sloping ledge; but the stream was then withdrawn from its roundabout course by a

short-cut, after which a sweeping meander wore out three short curves down-valley from the ledge; and still later, the next down-sweeping meander trimmed off the terrace front close to the ledge, preparatory to slipping past the obstacle and pushing back still farther the down-valley side of the terrace front. Another stage is

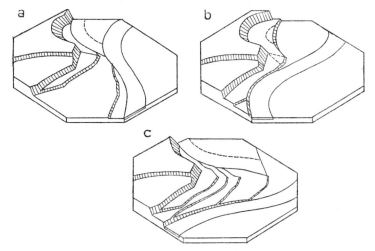

Fig. 2.18 The work of a compressed meander.

shown in Fig. 2.18 (*c*). Here the eighth westward swing of the stream is recorded. Part of the plain formed by the second swing happens to be still preserved, but none of the plains formed by the first and third. All the later swings (the fourth to eighth) are well indicated. A strongly compressed meander of the eighth swing has trimmed off all the terraces a little distance up-valley from their ledges, and would have trimmed them still closer but for being withdrawn by a short-cut. A later meander of the same swing is less successfully wearing away the down-valley extension of the terraces.

Diminished swinging of the meander belt

The greater the depth to which the valley floor is degraded, the more frequently may ledges be found, and, as a rule, the nearer they will stand to the axis of the valley. The number of defended cusps will therefore tend to increase as the valley deepens. The breadth of free swinging will at the same time decrease, and the space between the scarps of the lower terraces will necessarily be less than the space between the higher terraces. This principle, first stated by

54   *W. M. Davis*

Miller (1883), seems to be essential in explaining the stepping terraces of New England.

Distribution of high-scarp and low-scarp terraces

Ledges may gradually be disclosed at various points, each affecting cusp-making and terrace-keeping. The frequent swinging of the meander belt from side to side during the slow degradation of the valley floor requires that the discovery of every ledge lying well within the belt of wandering should be made soon after the stream has degraded the valley floor to the level of the ledge top.

This specialised conception of the terracing process leads to some reasonable deductions as to the distribution of high-scarp and low-scarp terraces. They are summarised in Fig. 2.19. A low-scarped terrace is formed near the far border shortly after degradation has

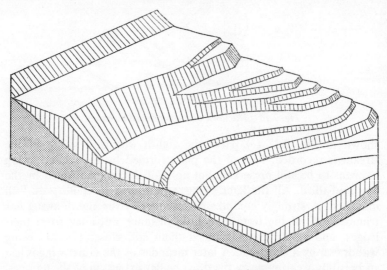

Fig. 2.19   The distribution of high- and low-scarp terraces.

begun. After six swings the river discovers a ledge somewhat within the belt of wandering. Then all the terraces lying behind this ledge will be preserved. On every later swing to this side, the river is halted nearer and nearer to the axis of the valley, and a flight of stepping terraces is formed in connection with a series of defended cusps. But on account of the absence of ledges on the near side, and of the increased breadth of wandering as the later stage of terracing is approached, the river destroys all traces of the earlier terraces in the foreground, where the tenth swing produces a single scarp by which the highest plain descends to river level. Then the eleventh and twelfth swings are held off from the high scarp by a lower ledge, on whose slope two low-scarped terraces are carved. It may be concluded that low, undefended, high-level terraces of early swings are most likely to be preserved behind defended cusps of later swings; that undefended terraces of early swings will probably be swept away, wherever broad swinging at low levels is not prevented; and that, when high scarps occur in a flight of stepping terraces, they are more likely to be found at or near the top than near the bottom of the flight.

Effect of rock barriers

Superposition upon strong rock barriers, which separates a valley into compartments, is a very familiar condition in New England. The present section considers the effect of rock barriers in producing fixed nodes, as Emerson (1898, p. 736) has called them, in streams which elsewhere vibrate freely. It is, however, not yet clear how a wandering stream will behave upstream and downstream of a fixed node. Several suppositions are possible:

1. The meanders will sweep down the meander belt, which will swing to and fro across the valley, but the amplitude of both movements will be decreased as the node is approached and extinguished as it is reached. So far as my observations go, this condition is more appropriate down-valley than up-valley from a fixed node. Below the node slight curves may be formed; these may develop into

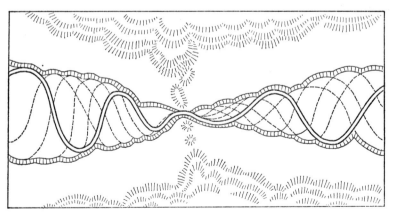

Fig. 2.20 Development of normal meanders below a node.

normal meanders (Fig. 2.20) as they sweep away from the sill; but such development will probably be gradual, and hence the valley floor will widen gradually in that direction.

2. The meanders may continue almost in full force as they approach the node from up the valley, merely changing in the lowest part of their course that leads directly to the sill. This might involve the introduction of a kink into the meander system, at the point of change from the normal downsweeping curve to the constrained course that leads to the ledge. That such a sharp bend is possible seems to be shown by certain peculiar forms in the meanders of the Theiss on the plain of Hungary, it being probable that bends of this kind result from the faster downsweeping of some meanders than of others. The considerable breadth of flood plain often observed next upstream from a node supports this supposition.

3. The fixed node may perhaps induce the formation of free nodes, evenly spaced from the ledge of superposition; then, between the fixed and the free nodes, the stream might vibrate as a stretched string does when it is lightly stopped at a third or a quarter of its length. Symmetrical free terrace cusps would result from this process; but so systematic a movement would seem to be possible only in rare cases, and further observation is needed to check if it can happen.

When two barriers occur near together, leaving a free space of half a mile or so, the river is fixed at two nodes but may vibrate between them. A remarkable case of this kind is found in the valley of Saxtons river at Bellows Falls, Vermont, as described on a subsequent page.

Relation of terrace patterns on the two sides of a valley

It is generally accepted, and repeatedly observed, that terraces on the two sides of a valley need not necessarily agree in number or in height. The relations of terrace patterns in plan have been less considered.

When a group of defended cusps occurs in a valley of moderate breadth, the stream must have been repeatedly deflected across the valley by the defending ledges, so as often to impinge upon the opposite side in about the same place. Hence re-entrants of more than usual size may there be worn out, next up-valley from which a group of free cusps may thus come to stand roughly opposite to the defended cusps. If the meander next above the ledge is somewhat compressed, the stream may strike more squarely across the valley and undercut the down-valley side of a terrace with somewhat greater vigour than usual. The valley of the Westfield river, a mile or so upstream from Westfield, Massachusetts, offers some remarkable examples of this kind.

When a side-stream enters the valley of a degrading main stream it tends to push the main stream away, causing it to wear out re-entrants opposite the confluence. In reacting from such re-entrants the main stream will strike across the valley, scouring out another group of re-entrants downstream of the confluence. When the deflection of the main stream is further guided by a ledge, the re-entrants will be all the more persistently and repeatedly carved out.

Ratio of sweeping, swinging and degrading

If numerous measures are taken of the difference of level between adjacent terraces in a certain section of a valley, it may be expected that two groups of minimum differences should be found — a group of smaller values representing the deepening of the valley floor in the interval between the down-valley sweeping of two successive meanders, and a group of larger values representing the deepening between two successive swings of the meander belt. If different measures of this kind were taken in different sections of a valley system, it might be possible to determine from their variations whether an even regional uplift or a tilting were chiefly responsible for the activity of the river in carving its terraces.

Assuming rapid uniform uplift, we should expect that the terrace scarp marking the interval between two lateral swings of the meander belt would be of greatest measure and relatively constant in the lower course of the main river, whereas a slow initiation of the uplift might possibly be recorded by a few low terraces at the top of the series. Slow close of the uplift and the very slow degradation of the valley floor in later time might be recorded by a few terraces of lower and lower scarps at the base of the series. If any terrace in the lower course of the river could be followed up the valley, it would assume a relatively higher and higher position in the series; for when the river had, in its lower course, worn down its valley floor to a low grade at the close of the period of uplift, there would still be degradation on the middle and upper reaches. So many, however, are the irregularities of drift terraces that, so far as I am aware, there has not been any systematic attempt to discover the facts by which these deductions might be confirmed.

Relation of the preceding deductions to the observations described in the following sections

The theory of terracing has here been presented before the observations of terraces are detailed, because it is the theory with its deduced consequences and not the facts that are on trial. Furthermore, it is only after the presentation of the theory that the pertinent facts can be selected. True, an attempt might be made independently of any theory to observe all facts thoroughly and to record them minutely, in the hope of including every item that could be asked for in the testing of whatever theory should afterwards be invented; but under this method of work items of minor importance are confused with those of major importance, so that the beginning is forgotten before the end is reached. Observational study of this kind is notoriously incomplete. The terrace problem, like many others, illustrates the difficulty – or impossibility – of seeing all the essential facts, and at the same time it illustrates the critical power given to observation, when this is directed towards significant points instead of being allowed to attempt the finding of all the facts before theorising is begun.

Largely deductive as the preceding portion of this essay is in its present form, the reader should not suppose that it was prepared independently of observation. The actual progress through the problem has involved repeated alternations of external and internal work; the collection of observations and the induction of generalisation on the one hand, and on the other hand the invention of hypotheses, the deduction of their consequences, the confrontation of deductions with generalisations, the evaluation of agreements and the repeated revision of the whole process.

IV. OBSERVATIONS OF RIVER TERRACES IN NEW ENGLAND

Valley of the Westfield river, Massachusetts: eastern section

This branch of the Connecticut rises among the Berkshire Hills of western Massachusetts, thence flowing eastwards part of the way across the broad valley lowland that has been excavated in the Triassic formation. Between the eastern base of the crystalline uplands and the ridge formed on the main sheet of Triassic extrusives, the stream has carved terraces in unconsolidated drift.

The village of Westfield, lying near the middle of the terrace system, marks the separation of its unlike eastern and western divisions. In the eastern division, Westfield river, reinforced by Little river, has opened a broad basin at an elevation of about 140 ft. The basin floor is nearly everywhere enclosed by a strong scarp of a single terrace whose plain stands at altitudes of 240 to 280 ft. The plain is not of simple origin. On the south-east its surface is rolling, as if consisting of morainic and kame-like deposits. On the north it is smooth, with sands fine enough to have been raised in occasional dunes; here the plain falls off south-westwards to the valley of Powdermill Brook in a series of lobes, whose intermediate depressions are too large to have been excavated by local drainage; hence it is probable that this part of the plain is a delta front in one of the areas of deposition described by Emerson (1898, pp. 650–3). South of the basin the smoother part of the high plain (Poverty Plains) is regarded by Diller (1877, p. 265) as an extension of the plain on the north, the originally continuous surface having been formed by the flooded Connecticut. Westwards up the Westfield valley the high plain ascends towards the hills and is of much coarser materials than elsewhere; this part seems to have been capped by local outwash. As the course upper gravels lie on fine sands and silts, this high plain is probably, like the one on the north, a delta surface built up in standing water.

The strong scarps B (Fig. 2.21) by which the high drift plains descend to the main low-level basin everywhere present concave re-entrants, whose curves unite in cusps – usually two-sweep cusps – showing that the streams have repeatedly swung against the scarps after the present grade was essentially reached. With certain exceptions, all these cusps – at least twenty-four in number – are undefended by ledges. A late stage of terracing has been reached, where the wide plain is nearly or quite reduced to grade with respect to a relatively permanent local base-level in the eastern notch in the ridge of extrusive rock. Many recent swings of the streams must on this account have tended to destroy earlier terraces by reducing them all to one level, instead of tending to make new ones at lower levels.

Westfield river is at present nowhere working against the base of the high terrace on the north, but several of its former courses along the terrace base are clearly revealed in shallow, swampy troughs. Little river was, in 1901, sweeping against its high terrace on the south at two points a little east of the railroad. Vegetation has been removed from the scarp, the sands are undercut, and the scarp face is sliding intermittently into the stream. Small dunes are formed at the top of the sliding bank by winds which sweep the sand up from below. It is evident that, by a repetition of sweeping and swinging of this kind, the high terrace has been worn back to its present outline.

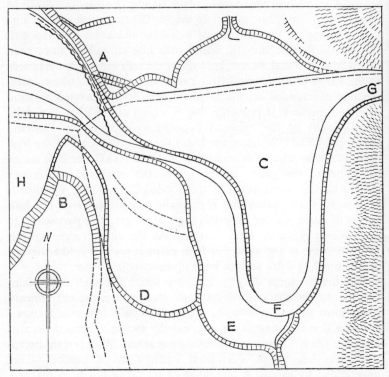

Fig. 2.21 Scarps by which high drift plains descend to main low-level basin.

Where the rivers have withdrawn from the high-scarped terrace, flat fans have been formed at the outlet of minor lateral valleys or beneath little gullies of wet-weather wash. The fan of Powdermill Brook, for example, forms a low barrier (X in Fig. 2.22) across a deserted channel of Westfield river, thus determining a swampy depression just north of Westfield Station. The further course of the brook follows the marshy deserted channels of Westfield river at the base of the scarp for over a mile.

It would be difficult to find better illustrations of the deductions previously offered. The two chief streams, far from exhibiting any incapacity to open their valley floors, have now widened them to a greater breadth than ever before. Whatever decrease of capacity may be due to decrease of stream volume and of stream slope, and whatever increase of work may be due to the more active wash of side-streams on account of gain in height of valley sides, the main streams are certainly more competent to corrade laterally than they have ever been, and there is every probability that they will in the

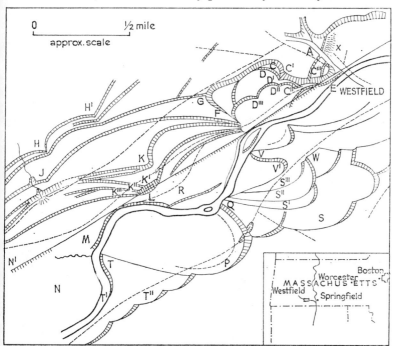

Fig. 2.22 Perspective diagram of the terraces of the Westfield river.

future continue to widen their basin still further by intermittent attacks upon its border until restrained by defending ledges or the hand of man. Indeeed, so nearly complete is the obliteration of all terraces above the level of the present basin floor, that one might be tempted to conclude that the Westfield and Little rivers never produced any series of flood plains in this division of their course at higher levels than those of modern times, until an examination of the western division of the Westfield terraces proves that flood plains must have been produced at various levels in the eastern division as well as elsewhere.

Evidently this example affords no indication that the production and preservation of terraces is due to any incompetence arising from decrease in the volume or other changes in the habits of New England streams. Terrace preservation must be due to some control external to the streams, and of this we find immediate proof on looking at the eastern and western enclosure of the broad basin just described.

The basin is enclosed on the east by the approach from the north of a defended spur (A in Fig. 2.21) towards a free spur, B, on the south; beyond the enclosure on the east a subordinate basin C is opened. The defended spur carries a terrace plain at a height of 200 ft, and the highest plain rises farther north by a faded scarp of gentle slope. Sandstone ledges are abundant along the western base of the spur; they are unusually steep, in part because of the eastward dip of the strata, and in part because of a certain amount of undercutting by the Westfield when it ran beneath them. The eastern side of the spur is not trimmed close to the defending ledges, but illustrates the unsymmetrical relationships shown in Fig. 2.17 (c). Widely as the river has swung from side to side in the basin farther west, it was here strongly constrained. Not only so; Westfield river has been somewhat impelled northwards by the entrance of Little river from the south (western side of Fig. 2.21), and it is probably in part at least on this account that the basin has been so well broadened northwards; yet on every sweep or swing against the sandstone reef the river was not only restrained from further northward encroachment at that point, but was deflected southwards across the valley. It is very probable that the excavation of the subordinate basin C is due to this cause, for it is opened farther to the south than to the north. Three strong southward loops of the river, including the present one (D, E, F) are here recorded, and it can hardly be by chance that the river has thus repeatedly turned southwards on its way to the fixed node G in the ridge of extrusive rock.

Nearly opposite this well-defended spur, but a little farther westward, the free spur B, rising to the full height of the drift plain, separates the subordinate eastern basin C from that part of the main basin H which has been scoured out by Little River. Unlike the defended spur on the north, the free spur is not a relatively permanent feature of the valley; it will be removed without difficulty if Little river sets about trimming away its western base. Nevertheless its occurrence today does not appear to be altogether a matter of chance, for it seems to illustrate the systematic features described earlier in the discussion of terraces on the two sides of a valley.

The main basin is enclosed on the north-west by a well-defended spur, Prospect Hill (*A* in Fig. 2.22), just west of Westfield Station. This will be further described with the terraces of the western division of the valley. On the south-east, Little river is held from swinging at present levels by superposition on a transverse sandstone ledge, to which brief reference will be made farther on. The contrast between the openness of the main basin, excavated where the streams have not been restrained by ledges, and the narrowness of the entering valleys where ledges have been encountered, is most striking.

Western section

The western division of the Westfield terraces, occupying the valley for about 4 miles from Westfield village to the base of the hills, is of greater interest than the eastern, inasmuch as it preserves the records of river work at many levels between the highest and lowest plains. The chief features of this locality are shown in a bird's-eye view in Fig. 2.22, as if looking north-east from a height of several thousand feet above the left front corner of the diagram. The railroad runs through the view for a distance of about a mile and a half. The foreground scale is larger than that of the background, and heights are exaggerated. Outcropping ledges are marked by hatching.

Between Westfield and the small rural settlement of Pochassic Street two miles to the west, many small ledges are exposed and many stepping terraces occur along the northern side of the valley. Few ledges are seen on the southern side, where the valley is generally bordered by a strong upper terrace with a few low terraces beneath it. On the northern side are four groups of defended terrace cusps, forming what may be called the Pochassic spur (just off the left of the diagram), Perry's spur (*K* in the diagram) and Prospect spur (*A*). Curved re-entrants have been excavated between the spurs where ledges are rare or wanting. The re-entrants show that the river has everywhere attempted to widen its valley, while the terraces on the defended spurs show that the widening has been locally prevented by the outcropping ledges. Wherever free cusps occur, they exhibit the patterns deduced as of common occurrence under the heading 'Ideal Terrace Pattern: Middle Stage'. None of the combinations deduced as rare is found. The cusps are usually more closely trimmed on the up-valley than on the down-valley side. It would be difficult to imagine a more complete confirmation of Miller's theory than is here presented.

Special mention may be made of a few features. Just east of Pochassic Street a series of nine terraces, H to M, may be counted. They range in height from 8 to 15 ft, and thus suggest a rough measure for the amount of valley deepening during a swing of the river southwards across the valley and back again. This maximum number is evidently dependent on the numerous ledges here discovered at all levels from highest to lowest. Although no other part of the valley shows so many terraces, it must be concluded that flood plains, continuous with the remnants here preserved, were made far up and down the valley; hence, that the river was essentially at grade during the whole process of valley degradation. Two terraces at the top of this flight, in the re-entrant east of Pochassic Street, exhibit minor re-entrants of small radius and large arc near H and H', comparable to the curves of the present river, indicating that no significant change of volume has occurred since the work of terracing began. A broad terrace plain stretches behind Perry's spur, K, with four low terraces rising above it to higher levels, showing that four northward swings were here executed. The fifth terrace, counting from the top of the series, runs forward to Perry's spur, because the highest ledge of that spur was discovered when the river was making its fifth northward swing. Several defending ledges in this spur would be unseen but for man-made cuttings. The fourth terrace swings forwards in a long sweeping curve to the apex of Brown's spur, F, because the summit ledge was there found by the fourth northward swing. Only two distinct terraces occur on the high plain behind Prospect spur, A, because the ledges in that spur rise still higher than they do in Brown's spur. The river has always shown a capacity for broad swinging until it became hampered by previously buried spurs. Brown's spur is peculiar in being closely trimmed on the down-valley side as well as on the up-valley side. Prospect spur has a terraced re-entrant, C, scoured out at mid-height with small radius and large arc, far back on its up-valley side. That is to say, a meander has there been twice compressed against defending ledges. Elsewhere the meanders seem to have slipped by.

The terraces on the south side of the valley are in several cases determined indirectly by the ledges on the north side. This is most distinctly the case where the river formerly swept forward from the lowest and farthest forward of the Pochassic ledges M, and consequently cut out one of the deepest re-entrants on the south side of the valley P. A single scarp now descends from the high-level plain into this strong recess. Similar but less manifest relations are suspected elsewhere; thus, K', K'', K''' on the north may correspond with S', S'', S''' on the south. Conversely, a number of low-level terraces remain on the south side of the valley south of Brown's spur, perhaps because the repeated northward swings of the river into the largest northward re-entrant, that between Brown's and Prospect spurs, have not required their removal. The numerous free cusps here found exhibit the features already deduced as of common occurrence. It is intended to make a close measurement of the slopes of these terrace plains in the hope of correlating the now separate remnants of single flood plains, and thus tracing the history of the terracing process in some detail.

Little river

A few words may be said about Little river, although the southern side of its valley has not been closely studied. The valley is divided into three sections by two barriers of sandstone, next upstream from which are considerable bodies of till. The till has been cut down to grade with the sandstone barriers, but the valley in the till is held to a small width, practically without terraces. Relatively few terraces are found even where the valley is bordered by stratified drift. The likely explanation is that Little river is smaller than Westfield river, and that a small stream must be hurried in attempting to keep pace with the degrading action of its master; it will have little opportunity for lateral swinging and terracing, except in one of two conditions.

The first of these obtains when the master stream has effectively ceased degrading its valley. This is now the case with the Westfield, because it has cut down upon a hard-rock barrier in the notch carved through the ridge of extrusives; it is probably for this reason that the lower section of Little river has swung broadly, opening the extensive valley floor which forms the southern part of the open basin east of Westfield. The enclosure on the south, where Little river is alone responsible for the form of its border, is nearly everywhere a single high-scarp terrace with numerous one-sweep or two-sweep cusps. Little river is swinging on its present flood plain more broadly than it has ever previously swung during the process of degradation.

The second condition obtains whenever the smaller stream becomes superposed upon a rock barrier; its work in the next upstream stretch then proceeds at its own rate, independently of the master stream. The valley floor in such a stretch tends to widen and thus to undercut all the narrower, earlier flood plains. This is the case both with the second and the third sections of the valley.

Valley of Saxtons river, Vermont

Saxtons river enters the Connecticut from the west at Bellows Falls, Vermont. Figs. 2.23 and 2.24 roughly illustrate its terraces here.

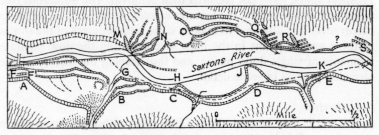

Fig. 2.23 Part of the Saxtons river valley, Vermont.

In the upstream of western section (Fig. 2.23) there are numerous ledges, but none of them has acted as a local base-level. The present valley floor is graded with respect to a heavy rock barrier a little east of the limit of Fig. 2.23, and at the western border of Fig. 2.24.

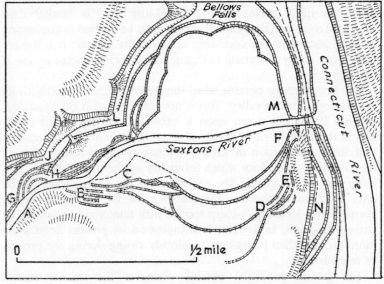

Fig. 2.24 Another part of the Saxtons river valley, beginning half a mile downstream from the reach shown in Fig. 2.23.

The stream has swung against the steep face of schist ledges at M, Q, and R, sweeping them practically free from drift on the up-valley side down to modern flood-plain level; but well-marked flights of terraces are preserved on the down-valley side of each ledge, where the trailing remnants of successive flood plains have been defended. At least ten different terrace levels can be counted adjoining ledge M. Their vertical interval, ranging from 5 to 10 ft, may again be taken to represent the amount of deepening between two northward swings of the stream. Ledges Q and R present similar features in flights of eight and six steps respectively.

Low-scarp terraces are wanting at high levels on the south side of the valley. The high upper plain descends by a single strong scarp, 20 ft or more in height. It presents a number of sweeping re-entrants between the defended cusps, A, B, C, D and E. The A–B re-entrant is floored by a rather uneven plain in which several indistinct terraces have been cut on what seems to be at least in part a mass of till, for large boulders are seen thereabout; and this plain is cut off in front by two terraces, whose blunt cusps from F to G appear to be determined in part by ledges and in part by boulders. The small tributary stream that crosses this re-entrant from the south has formed a fan on the high terrace plain and again on the floor of the re-entrant, but it is not dissecting the fans. No B–C re-entrant has been carved out, perhaps because till was discovered there. Several ledges were encountered at lower levels between G and H. A re-entrant was swept out between the defended cusps C and D when the river ran at a height about 10 ft over the modern flood plain, and another effort was here made to widen the valley floor at its present level; but as ledges are now discovered at H and J, farther forwards than C and D, the lower re-entrant has not quite consumed all of the earlier flood plain. A low terrace, caught on ledges J and K, stands in front of the re-entrant between D and E. The protection of the strong but low cusp at J as compared to that of the blunt but high cusp at D is one of the best illustrations of the effect of ledges in this valley. The river must have slipped past the ledge at D, as well as past most other defending ledges hereabouts, but a compressed meander must have been caught for a time on the ledge at J. Down-valley from E a modern swing of the stream has undercut all the earlier terraces, and a full-height scarp is the result.

It was on seeing, in October 1900, the relation of the defended cusps of the little terrace F to the corresponding defended cusp of the next higher terrace a little farther back at A, that the value of ledges in determining terrace pattern and in preserving the upper terraces from later attacks of the stream first came to my mind. The manner in which this explanatory ideal first took shape was as good an example of the sudden invention or birth of theory as I have ever experienced, for the theory was essentially complete at the moment of its first conscious appearance; since then it has only been confirmed by finding that it had already been anticipated by Miller.

In the eastern (downstream) section, Saxtons river illustrates terraces produced by a stream that has oscillated between two fixed nodes (Fig. 2.24). At the upper node the stream is narrowly held by ledges at A and G. A little farther upstream is a rocky gorge with cascades. The lower valley becomes somewhat more open as the space widens between the ledges B and C on the south, and J–H and L–K on the north. The small re-entrants between these ledges nearly everywhere bear the marks of having been energetically swept back at various levels. The stream has swung northwards at least nine times on the J–H group of ledges, and southwards at least seven times on the B group.

On leaving the cascade and the rapids below it, the stream has graded its course with respect to the eastern rock node between M and F–E; none of the ledges encountered on the way has done more than limit the breadth to which the successive flood plains have been opened. That degradation was gradual, giving the stream abundant time for broad swinging and wandering, is abundantly proved by the terrace remnants at various levels.

Passing the narrows at C, L–K, there is a broad stretch comparatively free from ledges, until the heavy ridge of rock, M, F–E, is encountered close to the junction of Saxtons river with the Connecticut. The ridge is now cut through by a narrow gorge, with falls on the downstream side. An oval plain, known as the Basin Farm, has been opened between the upper and lower narrows; at the level of its mid-height terraces the Basin plain probably had twice as great an area as it now has at flood plain level, but the reduction of area relates once again to the increasing constriction imposed by the mutual downstream approach of ledges.

The Connecticut below Bellows Falls, Vermont
The Connecticut river at Bellows Falls is superposed on a large body of rock, on whose down-valley side the river narrows in rushing cascades and rapids. A mile farther downstream three terraces are

developed (*A* in Fig. 2.25), on the west side of the valley just below the mouth of Saxtons river, *S*. All seem to be defended by ledges, the upper two by ledges of the large rock ridge at the mouth of Saxtons river (*N* in Fig. 2.24), the lowest by a ledge seen in the

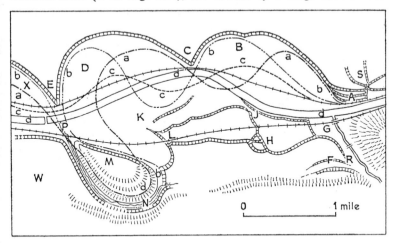

Fig. 2.25   The Connecticut river terraces from Bellows Falls to Walpole.

railroad cut a quarter of a mile south. A little farther south the two lower terraces are cut away westwards in the formation of a broad valley floor, *B*. A full-height scarp rises at the back border of the floor, with a blunt cusp near its middle determined by a strong ledge. The southern end of this open section is enclosed by a high free cusp, *C*, at whose apex the river is now working. Another open valley floor, *D*, follows the free cusp and is limited by a second free cusp, *E*, of less height but of greater forward reach than the first. The meaning of these two free cusps will be considered below.

On the east side of the valley, past the entrance of Cold river (*R*), ledges are found to be more numerous. A broad mid-height terrace was opened until a ledge, *F*, was discovered in the base of the uppermost terrace. The mid-height terrace was cut back by a much later swing of the river near present flood plain level, until a high ledge, *G*, was discovered. Nearly a mile farther south is a group of defended terrace cusps, *H*, up-valley from which the river has swept out some vigorous curves. Near this point the lower terrace advances to the river bank on account of the farther forward reach of other ledges, *J*; one of them now outcrops in the river bank. It is evident that the meander belt of the river has here been constrained to take a more and more westward course as it cut deeper and deeper, and it is probably on this account that the first large western re-entrant below Saxtons river has been so thoroughly scoured out at a low level.

The lower eastern terrace is gradually cut back down-valley from the foremost defending ledge, a broad low plain, K, being opened to a half-mile width, after which it narrows towards the bridge between Walpole (W) and Westminster (X). The mid-height terrace continues down-valley, first showing an apparently free two-sweep cusp, then a defended cusp, L, the defending ledge at the base of the latter being disclosed in a railroad cut. After this the terrace is cut some distance back in a low-level re-entrant, enclosed by a scarp concave southwards (down-valley). A little farther on, an isolated hill, M, is separated from the eastern side of the valley by a deep trench, N, of large sweeping curvature to the north-east, and for the most part apparently cut in till. The trench has a rather strong undercut slope on the outer side of its curved course, and a gently terraced slope on its inner side. It marks a former path of the river around a lobate spur; the river was diverted at a comparatively modern date by wearing through the narrow neck of the spur, P, a little upstream from Walpole Bridge. The second, E, of the two free cusps on the west of the river is the unconsumed remnant of the neck of the spur, west of the cut-off.

The following explanation of the relation between the two free cusps C and E may be suggested. At an early stage of the time during which the river was making its great north-eastward detour around the spur EM, its course may be represented by the curve $a \ldots a \ldots a \ldots a$. The normal order of change in these curves would develop a later course $b \ldots b \ldots b \ldots b$, thus opening out the two large westward re-entrants B and D, leaving the free cusp C as yet unconsumed between them. Had this process continued, the free cusp C would have been worn away by the first sweeping of the $b \ldots b$ meander; but before this was accomplished, cut-off occurred at P.

Now the detailed maps of the Mississippi River Commission show that when cut-off occurs a systematic series of changes is initiated, and is extended both upstream and downstream from the cut-off. The straightening of the course at the cut-off tends to straighten it elsewhere also; the tendency is, however, most active near the cut-off and weakens with distance from it. Accordingly, the course $b \ldots b \ldots b \ldots b$ is reconstructed as changing to $c \ldots c \ldots c$

River Terraces in New England 71

... *c*, and then to *d* ... *d* ... *d*, the line of the present channel. The river is thus withdrawn from both of the westward re-entrants, around whose up-valley curves it was probably flowing when the cut-off took place. Thus the free cusp *C* is left unconsumed, for a time at least. It may perhaps be possible, by accumulating additional examples, to give some degree of verity to this rather hazardous explanation.

The Connecticut below Turners Falls, Massachusetts
For several miles down the river from Turners Falls to the railroad bridge in Montague, there are numerous examples of defended terraces, fully confirming the principles already illustrated. Concerning a stretch of river southwards from this section, Emerson (1898, p. 725) has written:

> The subsidence of the waters of the Connecticut lakes to the present Connecticut river was very rapid. ... As a result, one goes down – through the whole length of the Montague Lake, which was well filled up in the flood time, except in its southern portion – by a great scarp to the series of erosion terraces of the modern river, the highest of which rise but a few feet above the level of the flood plain.

This conclusion as to the 'very rapid' change of level, by which erosion of the present valley floor was initiated, seems to be based entirely on the great scarp descending from the high drift plain to the low terraces of the modern river. This conclusion, so directly opposed to the inferences presented here, appears of doubtful validity, for reasons now to be stated:

1. The single great scarp does not necessarily prove that the river suddenly cut its channel down from the level of the 'lakes' to about that of the modern flood plain. The single scarp is perfectly consistent with leisurely degradation, and with the production of numerous flood plains, provided only that the modern swinging of the river has been greater than former swinging at higher levels.

2. The leisurely lateral swinging of the Saxtons and Westfield rivers at high levels, as recorded by the upper members of the terrace flights in their valleys, show that they were not hurried in the early stages of their work of degradation; yet hurried they would surely have been, had the master river which they join suddenly entrenched itself into the weak drift fill of its aggraded valley. There are, to be sure, certain rock and till barriers between these terrace flights and the confluences with the Connecticut, and

such barriers might separate a quickly degrading trunk river from slowly degrading tributaries; but it is believed that the barriers are too low to have been encountered until after the high-level terraces had been carved.

3. There are certain points where the Connecticut itself exhibits stepping terraces at altitudes of at least 8 ft or more above its present level. The best of these are at East Deerfield, 2 miles south of Turners Falls. For the heights of five successive scarps on the west side, south of the railroad station, I estimate the following values (in feet, in descending order): 30, 25, 15, 18, 35 – total, 123 ft. A run northwards instead of eastwards includes seven scarps, estimated at 30, 25, 10, 5, 15, 10 and 15 ft – a total of 110 ft. A leisurely process of degradation, with repeated lateral swinging, is once again probable.

REFERENCES

ADAMS, C. B. (1846) *Second Annual Report on the Geology of the State of Vermont.*
DILLER, J. S. (1877) 'Westfield during the Champlain period', *Amer. J. Sci.*, XIII 262–5.
EMERSON, B. K. (1898) *Geology of Old Hampshire County, Massachusetts* (Washington, D.C., U.S. Geol. Survey) Monograph 29.
HITCHCOCK, E. (1857) *Illustrations of Surface Geology* (Washington, D.C., Smithsonian Institution) 155 pp.
MILLER, H. (1883) 'River terracing: its methods and their results', *Proc. Roy. Phys. Soc. Edinburgh*, VII 263–305.

3 Longitudinal Profiles of the Upper Towy Drainage System

O. T. JONES

THE High Plateau of central Wales has been deeply dissected by numerous rivers, some of which, such as the Severn, the Wye and their tributaries, flow eastwards, ultimately reaching the Bristol Channel; others, such as the Rheidol and the Ystwyth, flow westwards across the Coastal Plateau into Cardigan Bay. The Teifi, which reaches that bay much farther south, lies mainly on the Lower Plateau, its course being for some distance almost parallel to the boundary between the Coastal and the High Plateau. The Towy with its main tributaries arises near the summit of the High Plateau, and discharges into Carmarthen Bay. The southward-flowing trunk stream receives in succession the Camddwr and the Doethie-Pysgotwr, which come in from the north-west (Fig. 3.1). South-west of the Pysgotwr the Cothi flows for a few miles towards the Towy, then turns abruptly at right-angles, eventually joining the Towy at Nantgaredig many miles downstream. The portion which may be distinguished as the Upper Cothi formerly continued south-eastwards along the Gwenffrwd valley into the Towy, but has been diverted into its present course by capture. The Camddwr and Doethie-Pysgotwr join the main river above Rhandirmwyn, which is a mining village 7 miles north of Llandovery and about 15 miles from the source of the Towy. In the 25 miles which intervene between this place and Carmarthen, where it becomes tidal, the Towy receives only four tributaries comparable in size with those mentioned above, namely, the Brân at Llandovery, the Sawdde at Llangadock, the Cothi at Nantgaredig and the Gwili at Abergwili.

Above Rhandirmwyn there is a remarkable contrast between the upper and the lower portion of each of the valleys, and if the longitudinal profiles are plotted from the data on the 1-in. Ordnance Survey map, a marked discontinuity of gradient in the middle portion of each valley becomes evident, the contrast between the immature or youthful character of the lower reaches of the valleys and the mature forms of their upper reaches being clearly seen on the profiles. As the tributaries of the Towy below

Rhandirmwyn do not apparently reveal these features, the explanation therefore must be sought for above that place.

Preliminary observations many years ago were followed in 1920 by the theodolite determination of a large number of levels in the

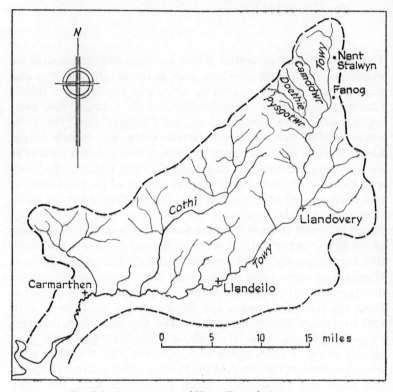

Fig. 3.1 Location map of Upper Towy drainage system.

valleys, and in 1921 by additional levelling. The present communication embodies the results of these surveys.

I. BASE-LEVELS

The existing Towy valley as far up as the neighbourhood of Fanog has been excavated in the floor of an earlier valley. Where the river passes from the old valley into the new, a change in valley-side slopes and in stream gradient sets in, so abrupt that it may be called a topographic unconformity. The gradient near the source is nearly 250 ft per mile, diminishing to about 40 ft per mile above Fanog. Here it suddenly increases to something like 200 ft per mile, thereafter diminishing continuously to about 2 ft per mile in the tidal region at Carmarthen. Similarly, the relatively gentle slopes of the old valley give way suddenly to the precipitous slopes of a ravine, and slopes of the same inclination as those of the old valley do not reappear for some distance downstream.

The discontinuity in the topographic features of the valley shows most clearly in the longitudinal profile. In the transverse profile it is more rapidly obliterated, so that a few miles below Fanog it is difficult to discover that the valley-side slopes have been refashioned.

Generally similar relationships, with certain variations, obtain in the valleys of the Camddwr, Pysgotwr and Doethie; the discontinuity is also well marked in the Cothi valley, although it is there complicated by stream derangement.

The change in each valley can be attributed to rejuvenation, resulting from elevation which greatly increased stream gradients and which led to renewed erosion. Between Carmarthen and Fanog the Towy is graded to present sea-level, and the tributaries which enter it are at grade with the existing river; but from Llandovery to Fanog the graded condition has not been fully attained. The valley above Fanog is not graded with that below. It appears to have been fashioned at a time when the level of the land stood relatively lower than at present. The tributaries which enter above the rejuvenation head are, with minor exceptions, approximately graded to the old valley, whereas below the rejuvenation head only the upper part of each tributary valley above its own rejuvenation head appears to have been graded to the old Towy valley.

In this region, that is to say, we have evidence of two distinct base-levels which correspond to two different cycles of erosion. The lower portion of each valley has been determined by the existing level of the sea as a base-level, while the upper portions of the valleys are related to a different and older base-level. These levels will be referred to as the Llandovery and Fanog base-levels respectively.

As will be shown, there is also some reason to infer a third base-level, to which the uppermost regions of the valleys correspond. This will be referred to as the Nant Stalwyn base-level.

II. LONGITUDINAL PROFILES

The Towy

The profile obtained by plotting the levels of points on the floor of the valley against the distance from the source measured along the mean course of the valley is a rather irregular curve, in which short reaches of relatively high gradient alternate with longer reaches of lower gradient; but despite numerous irregularities it is obvious at a glance that the average gradient diminishes downstream, since a smooth line drawn through the levelled points so as to represent the average level of the valley floor is a concave curve which steepens rapidly towards the source of the river. The average gradient at any point in the valley can be determined by estimating from this curve the level of the floor at numerous points – say, at each fifth mile – and dividing the difference between adjoining points by the distance between them.

If the values of the gradient so obtained are also plotted, and a smoothed curve drawn through the points, the change in the gradient is strikingly shown. It diminishes rapidly in the first 2 miles of the valley, and thereafter much more slowly.

Considerable differences are revealed between the actual level of the floor and the average profile represented by the smoothed curve. The reason for these differences need not be considered at this stage, but it is obvious that the river is not in complete adjustment with the valley floor. In those parts of the valley where the floor is above the average level, the river is confined in a narrow rock channel where active erosion is in progress. In the intervening reaches its channel lies in gravel, and is bordered by a varying width of alluvial deposits, so that the river there has the appearance of being fully graded. Each rock barrier serves as a temporary base-level for the graded reach immediately above, and with its progressive lowering the grading of the upstream reach keeps pace with the changing base-level. The profile is therefore approaching closely to the graded condition, and when the river has worn down its rock channels by a few feet, the level of the valley floor will not depart sensibly from a smooth curve, drawn so as to pass through the majority of the points in the existing graded reaches. The amount by which the rock channels must be lowered to attain this profile varies from 4 to 12 ft in the lower region (between 4 and 8 miles) and from 14 to 22 ft in parts of the upper region (between 2 and 4 miles).

As regards the lower region, it is probable that in pre-Glacial times a mature profile had been attained, and that the existing irregularities have been caused in various ways by the subsequent occupation of the valley by moving ice; but as regards the upper region, especially between 3 and 4 miles, there was probably an abrupt change in the level of the valley floor, which has been partly reduced or toned down by glacial erosion. The pre-Glacial profile of the Upper Towy seems in fact to have consisted of two mature reaches, separated by a step in the valley floor which was in process of being lowered by river erosion. These two reaches correspond to two periods of base-levelling. The irregularities in the longitudinal profile are, however, of minor importance in comparison with the size of the valley, which on the whole presents a thoroughly mature aspect with relatively smooth convex slopes. We may therefore assume that the smooth curve drawn through the graded reaches represents the form of the mature valley that will in time be excavated in the slaty rocks by the Towy. Such a curve gives a closer approximation to the fully graded profile of the valley than one which represents the average level of the valley floor. It is possible, however, that in the fully mature stage the profile would be slightly more concave than this curve. This is suggested especially by the fact that the gradients between 4 and 6 miles are, on the whole, in excess of a smoothed gradient curve (cf. Fig. 3.2). The slopes of all the tributary valleys are similar in form to those of the main valley, and their floors have been eroded to a concave profile, so that we are clearly dealing with a drainage system in a mature stage. The mature valley persists down to the point where active erosion of the floor and slopes is in progress. Even below that point, in most of the valleys, there are more or less obvious traces of the pre-existing valley floors. As there is no reason to believe that the profiles of the valleys above their points of rejuvenation have been altered by the subsequent erosion of the valley below, the mature valleys must have existed substantially in their present form since the time when they were graded during a former period of base-levelling. It is probable, however, as suggested above, that some modification of their floors and slopes may have been produced by glacial erosion.

The smooth curves obtained in the manner described above, which represent (*a*) the level of the valley floor at increasing distances from the source, and (*b*) the gradient of the valley, indicate that the levels and gradients steadily diminish downstream, or (in other words) that the valley becomes lower and markedly flatter in that direction.

The detailed form of these curves is probably determined for each river mainly by the volume of the river during flood. The nature

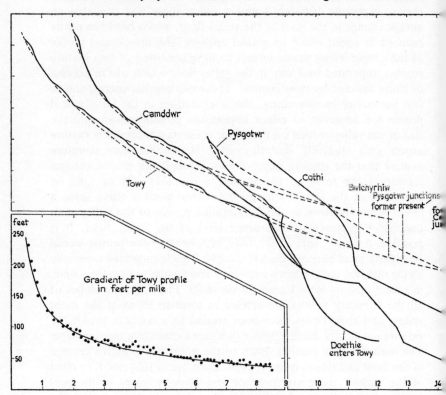

Fig. 3.2 Longitudinal profiles (i).

of the channel, or of the deposits in which it is excavated, may also exercise some influence. These deposits vary in size from large boulders down to fine mud, and the form of the longitudinal profile reveals the gradient required by each river to transport its normal load. The volume of a given river depends upon the area of the catchment, the rainfall and the run-off; the flattening of the profile downstream is undoubtedly related to the increase of volume which goes with increase of the drainage area.

It is important to bear in mind that the profiles that we have been discussing are those of the existing rivers in relation to their present circumstances; but if next we proceed on the hypothesis that the ancient Towy river which excavated the mature pre-Glacial valley had the same profile as the existing river, we can, by extrapolation, reconstruct the ancient valley in the region below the point where it has been destroyed by rejuvenation, and also obtain an estimate of the present height above sea-level of the ancient base-level. It cannot, however, be assumed without some knowledge of their regime that the ancient rivers had the same profile as their present-day successors, since their volume and load may have been different. The influence of these factors is discussed below.

In the absence of information regarding the form of the rock floor in pre-Glacial times, the relation which the profiles of the existing rivers bear to those of their predecessors cannot be accurately determined. We find, however, that each river encounters rock in innumerable places, not only at the side of the valley but in the centre as well, which may be interpreted to mean that the cover of superficial deposits is in general of limited thickness, and that the present profile of the rock floor coincides approximately with the profile of the flood plain. Further, there is no reason to believe that the glacial excavation was so severe as to invalidate the assumption that the profile of the ancient valley was not materially different from that of the recent valley. It is true that in parts of the valley there are rock steps which indicate inequalities in the floor, but in comparison with the width and depth of the valley they are of inconsiderable dimensions.

If we can succeed in finding a mathematical expression to which the profiles of the existing rivers conform, those profiles can be prolonged by extrapolation and thus afford a means of reconstructing the floors of the ancient valleys which have been destroyed by erosion. Moreover, the base-level to which the ancient drainage system was graded can be approximately determined. In applying this method we can make use of the principle that in a mature system the level of the flood plains of a tributary and of a main river are approximately the same at their junction. The rock floors are not in general accordant, because the depth of river deposits in the main valley is usually greater than in the tributary valley; the discordance is, however, comparatively small, and corresponds to the difference in the rise of the two rivers during floods.

The extrapolated profile of a tributary should therefore give nearly the same level for the ancient valley floor at the junction as does that of the main river. There is probably also some relation between the gradient of a tributary and that of the main river at the point where they meet, but the nature of this relation has not, so far as I know, been determined. If the law connecting the gradients were known, a further important principle would be available for the reconstruction of the old valleys. The appearance of the profiles published by Penck (1894, p. 323, Fig. 23), de la Noë and de Margerie (1888, Plate 18), and others suggests that the profile of a main river is an envelope to the profiles of tributaries, in which case all the curves should have a common tangent at their junction, or (in other words) should have the same gradient at these points. This is almost certainly not the true relation; rather there appears to be some evidence of a discontinuity at the junction of a tributary with the main river and that the gradient of the former is, in general, higher than that of the latter, the difference increasing with the difference between the volumes of the rivers. This is more evident in small tributaries than in large; the alluvial cone of a small stream debouching on a wide valley floor has usually a well-defined margin, and rises sharply out of the flood plain of the main river.

The appearance of the profile suggests that it might be expressed by a parabolic, a hyperbolic or a logarithmic function, and various forms of all these functions have been tried. The best expression appears to be that which was suggested by the form of the gradient curve, which resembles a rectangular hyperbola. This curve can be expressed by the formula $(x+a)(g+b) = k'$, where x is the distance measured from the arbitary origin of the river, g is the gradient dxy/dx, and a, b and k' are constants (a and b may be positive or negative). It is found that a formula of this type in fact gives a fairly close representation of the gradient in different parts of the valley. If the expression be integrated, we obtain an expression of the following form for the profile: $y = k \log(x+a) - b(x+a) + c$, where x, a and b have the same significance as above, y is the height of the valley floor above Ordnance Datum, and k and c are new constants. This represents a logarithmic curve on which is superposed a straight line, or, in other words, a logarithmic curve referred, not to a horizontal, but to an inclined datum line. The logarithmic component is the more important for moderate distances from the origin. The corresponding gradient, g or dy/dx, is given by the expression $g = [k'/(x+a)] - b$ where $k' = k \log_{10}e = 0.434k$ or $(g+b)(x+a) = k'$. Since the value of y diminishes with increasing values of x, the gradient is negative, and since $x+a$ is positive, then

Longitudinal Profiles of the Upper Towy Drainage System   81

k' and k must be negative. The constant $c = Y - b$, Y being the value of y when $x + a = 1$; but since b is in general small, c is approximately equal to Y. The expression is best used for purposes of calculation in the following form:

$$y = c - k\log(x+a) + b(x+a).$$

The disadvantage of the expression is that, unless b is zero, the gradient does not vanish for infinite values of x. This would appear to be a necessary condition, since the main river ultimately flows into the sea where the gradient is sensibly zero; but if, as suggested above, there is a discontinuity of gradient where two bodies of flowing water unite, the smaller body may retain a perceptible gradient up to the point where it becomes merged in the larger body. Another disadvantage is that the value of the constants can only be obtained by trial. This has been done by first constructing a smoothed gradient curve, and, on the assumption that it is a rectangular hyperbola, finding an approximate value of a from this curve. The other constants corresponding to that value of a, and also other values differing from it by small amounts, were then determined by calculation. For each set of constants so obtained the values of y for a number of points along the valley floor were calculated, and compared with those read off directly from the curve. That value of a which gave the closest agreement, especially in the lower part of the profile, was selected for the purpose of extrapolation.

If b is assumed to be zero, an approximate value of a can be obtained directly, and the substitution of this value of the constant in the formula gives approximate agreement with the curve; but a better result is, in general, obtained when a slightly different value of a is used. The gradient curve was constructed by determining from the smoothed profile the level of the valley at each tenth mile, taking the difference between alternate pairs corresponding to a distance of one-fifth of a mile, and plotting the resultant gradient against the values of x corresponding to the middle of each interval. A smoothed curve was then drawn through the points so obtained (see Fig. 3.2).

The three points $x = 0\cdot5$, $4\cdot5$ and $8\cdot5$, measured in miles from the arbitrary origin selected, were used in finding the constants. The constant b is zero for a value of $a = 1\cdot86$, but a closer correspondence with the smoothed profile is obtained when a is somewhat greater than this figure. Table 3.1 shows the values of c, k and b in the above expression for varying values of a. Table 3.2 is a summary of the results obtained by the use of various constants; it shows the value of y corresponding to the mile-points between 1 and 8 miles from the origin, and the difference between the calculated value for

TABLE 3.1

Values of the constants c, k and b for different values of a in the expressions

$$y = c + k \log (x+a) + b(x+a) \text{ and } g = -\frac{dy}{dx} = -\frac{k'}{x+a} - b$$

a	0·6	0·9	1·2	1·8	2·0	2·4
c	1462	1522·3	1590·6	1742·5	1797·8	1915·8
k	−487·6	−578·8	−679·1	−892·4	−967·6	−1124·4
b	−17·02	−12·66	−8·3	+0·13	+2·88	+8·32
k'	−211·78	−251·39	−294·24	−387·57	−420·23	−488·33

each point and that obtained from the curve. The calculated levels of the reconstructed Towy valley at the points where the Camddwr and Pysgotwr enter, and also at the mouth of the present river below Carmarthen, are also given. The level of the old valley at its junction with the Pysgotwr is calculated on the assumption that the two rivers met above the Ystradffin gorge. The level at the junction has also been calculated on the assumption that they met at Towy Rocks, as they do at present. For comparison with the tributaries, the gradient of the Towy at its junction with the Pysgotwr on the first of these assumptions is shown.

It will be seen from the table that, while all values of a give levels of the valley floor between the third and the eighth mile that differ by only a few feet from the curve and in some cases differ only by inches, no value gives a satisfactory agreement at the first mile and only the smaller values at the second mile. Greater importance is attached, however, to close agreements with the lower part of the curve, since our confidence in reconstructing the old valley by extrapolation is increased in proportion to the degree of accordance between the calculated and the smoothed profile in the lower region. In this respect there is little to choose between the values of a in the last three columns, for each of which the average difference lies between 0·3 and 0·5 ft, which are within the limits of error of plotting of the results, and the individual differences are somewhat irregular, some values being too large, others too small.

From these results we find that the level of the ancient valley

Longitudinal Profiles of the Upper Towy Drainage System

TABLE 3.2

y calculated for different values of a and differences from curve

x	y obtained from curve	0.6	diff.	0.9	diff.	1.2	diff.	1.8	diff.	2.0	diff.	2.4	diff.
1	1331	1335.3	4.3	1336.9	5.9	1339.8	8.8	1344.8	13.8	1344.8	13.8	1346.5	15.5
2	1215.5	1215.4	−0.1	1217.9	2.4	1221.0	5.5	1225.6	10.1	1226.8	11.3	1228.9	13.4
3	1134	1129.4	−4.6	1130.8	−3.2	1132.5	−1.5	1135.2	1.2	1135.9	1.9	1137.2	3.2
4	1063	1060.5	−2.5	1060.7	−2.3	1061.2	−1.8	1061.9	−1.1	1062.2	−0.8	1062.6	−0.4
5	1000	1001.9	1.9	1001.4	1.4	1001.0	1.0	1000.4	0.4	1000.3	0.3	1000.0	0.0
6	917	950.1	3.1	949.4	2.4	948.6	1.6	947.4	0.4	947.1	0.1	946.4	−0.6
7	901	903	2.0	902.7	1.7	901.9	0.9	900.7	−0.3	900.5	−0.5	899.8	−1.2
8	859	859.9	0.9	860.1	1.1	859.7	0.7	859.2	0.2	859.1	0.1	858.8	−0.2
9.6 (Camddwr)		796.6		798.3		799.2		800.8		801.3		802.2	
12.3 (Pysgotwr)		700.8		706.6		711.0		718.7		723.7		725.6	
56 Mouth of Towy		−352		−214		−77.6		177.6		258.8		415.6	
Sum of differences		19.4		20.4		21.8		27.5		28.8		34.5	
Sum of differences between 5 and 8		7.9		6.6		4.2		1.3		1.0		2.0	
Gradient at the mouth of the Pysgotwr	33.5		31.8		30.2		27.4		26.6		24.9		
Levels in feet if the rivers met at Towy Rocks	684		691		696		705.5		708		713.5		

floor at its junction with the Camddwr varies between 796·5 and 802·2, the most probable value being just above 800 ft. This confirms the interpretation of the shelf near the junction of these valleys which stands at 800 ft above Ordnance Datum as a remnant of the old floor. The level of the present valley floor at that point is 685 ft, or 115 ft lower. Similarly, there is no great difference between the extreme values for the level of the Towy–Pysgotwr junction, which range from 700·8 to 725·6. We may take 723·7 or 725·6, namely those given by the curves for which $a=2·0$ or $a=2·4$, as being nearest the true values. The existing valley floor is 265 ft lower, that is, about 460 ft above Ordnance Datum. The difference between the old valley floor and the present valley thus appears to increase at the rate of about 50 ft per mile, but it is evident that it cannot continue to increase at this rate. On the other hand the difference will not become smaller downstream unless, as is unlikely, the gradient of the old floor was steeper in that direction than that of the present floor. A rough approximation to the base-level of the old valley system is thus obtained. It now stands at least 265 ft higher than the present sea-level.

If we attempt to determine by extrapolation the values of y at the present mouth of the Towy, 56 miles from the origin, slight differences in the value of a give rise to wide divergences in the level of the old floor at that point. Thus, for $a=1·8$, $y=178$; for $a=2$, $y=259$; and for $a=2·4$, $y=415·6$. The first two figures are (as shown above) too low, and the last may be too high, but rough limits are thus obtained.

A closer approximation to the ancient base-level can be obtained by making use of the profile of the present river as a control in extrapolating from the upper part of the curve. Between Rhandirmwyn and Llandovery the Towy is, in places, confined in a narrow rock gorge. At those points active erosion is in progress, and the profile has not acquired its final form; but between Llandovery and Carmarthen the river meanders through a wide alluvial plain, and its profile has attained stability.

The values of y along the ancient valley have been calculated for $a=2$ and $a=2·4$ from the head of rejuvenation above Fanog to the mouth of the Towy 6 miles below Carmarthen, and the profile of the present valley floor between the Towy–Pysgotwr junction and the same point has been constructed from the contour lines and spot-levels on the 1-in. maps. The flood plain at the mouth of the river is about 13 ft above Ordnance Datum. The latter profile was copied on tracing-paper, and compared by superposition with the two calculated profiles, the base lines being kept parallel. The curve for $a = 2·4$ coincides with the recent profile between 30 and 38 miles; above 30 miles the recent valley floor has a higher gradient, as might be expected, since in that region it is in process of being lowered by erosion. Below 38 miles also the recent valley has a slightly higher gradient than the curve, but if we assume that the calculated curve is not seriously in error at 30 miles, and that thereafter it has the same profile as the present floor, it would stand at the mouth of the Towy at 397 ft above Ordnance Datum. A well-graded river traversing an area deeply covered by the products of rock decay would probably have a flatter profile than the present Towy, wherefore this figure may be too low. The value of y at that point obtained by calculation from the formula is 415·6, corresponding to an ancient 'Ordnance Datum' of about 403 ft.

With the value of $a = 2$, coincidence between the two profiles is obtained between 22 and 28 miles, but the fit is evidently less satisfactory than with the previous curve, especially below 28, where the recent valley has considerably the lower gradient. Assuming that the calculated profile is approximately correct at 22 miles, and is replaced below that point by the recent profile, the ancient valley floor at the mouth of the Towy would stand at 350 ft above Ordnance Datum; the value given by the curve is 259 ft. These values are certainly too low, those obtained from the previous curve being probably nearer the truth.

We thus arrive at the conclusion that the ancient base-level is now at least 384 ft, and possibly as much as 403 ft, above its former level.

We arrive at somewhat similar figures by using the recent profile as a control in another way. After the gradient of a mature valley has attained a given value, it diminishes downstream according to some law that is characteristic of the river. This law may be expressed approximately, as shown above, by a formula, but as there are points in the present valley (say, near Rhandirmwyn) which have the same gradient as the old valley above Fanog, the law of change of gradient may also be obtained by superposing a part of the recent profile on some part of the profile above the head of rejuvenation, so that they coincide for a certain distance (the base lines of the two profiles being kept parallel as before). There is, therefore, no extrapolation of the ancient valley, since the prolongation downstream of the profiles of the old valley is assumed to be the same as that of the recent profile. The result is interesting: it is found that, by superposing point 12·8 of the recent profile on 8·2 of the ancient profile – that is, a third of a mile above where the ancient valley is being eroded away – the profile calculated from the formula in which $a = 2·4$ does not depart anywhere from the recent profile by more than 10 ft between 8·2 miles and Carmarthen, a distance of about 40 miles. The average departure is $3\frac{1}{2}$ ft. The formula gives, therefore, an almost exact representation of the form of the recent profile for a distance of over 40 miles, and it is clear that (although it is based on information derived from points only a few miles from the source) it summarises the behaviour of the river many miles below with some exactness. In order to produce the superposition of the curves, the base or Ordnance Datum line of the recent profile has to be raised 414 ft, and this figure gives another estimate of the difference between the ancient and recent sea-levels. The superposition has been done in another way by making use of the gradients. Those of the old valley were calculated from the formula for two values of a (2 and 2·4) and plotted against the distance from the source. The curves so obtained were traced and superposed on the gradient curve of the recent profile, so as to bring some one point where the gradients were the same into coincidence. If the two curves then fitted closely, it followed that the gradients at the other points downstream were the same, or, in other words, the two profiles had the same form. It was found that, by bringing the gradient corresponding to point 10·5 on the old profile ($a = 2·4$) to coincide with that at point 15·2 on the recent profile, the two curves fitted closely. The level at point 10·5 is 774 and at 15·2 it is 362, the difference being 412, agreeing very nearly with the previous result. The last two values are believed to be somewhat too high, as, by the method of superposition employed, a part of the river which has a large volume is superposed on a part which has a smaller volume. For example, the gradient at the lower point 15·2 will ultimately be lower than it is at present when the valley above Llandovery has been more perfectly graded, so that, in order to obtain the same gradient as at 10·5, a part of the recent profile lying higher upstream (and, therefore, at a greater height above Ordnance Datum) would have to be superposed on the ancient profile, and a smaller difference between the levels at that point would result.

If the same process is attempted for the curve in which $a = 2$, it is impossible to obtain coincidence of either the profile or the

Longitudinal Profiles of the Upper Towy Drainage System 87

gradient curves for more than a short distance, and it is clear that this curve represents much less closely the characteristic behaviour of the river than the curve obtained from the higher value of a.

Summarising these results, we find that the base-level to which the ancient Towy drainage system was graded now stands at a height of between 384 and 414 ft above the present sea-level. The latter value is probably too high, and the former possibly too low, but these studies have made it possible to define the change in base-level within very narrow limits (Figs. 3.2, 3.3).

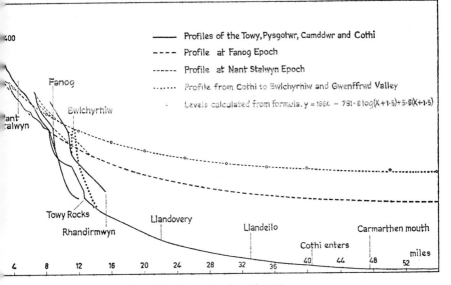

Fig. 3.3 *Longitudinal profiles (ii).*

The argument rests, however, on two important assumptions: (1) that the ancient river was comparable in (a) its volume and (b) the nature of its load with the present river; and (2) that there has been no tilting of the area subsequent to the grading of the mature valleys which are now preserved in the Upper Towy system. There is abundant evidence that the ancient valleys had been destroyed by erosion for a distance of nearly 50 miles above the mouth of the present Towy before the advent of the Glacial epoch, and that the displacement of the point of rejuvenation upstream by Glacial or post-Glacial erosion has been relatively trifling. The change of base-level which caused the rejuvenation must, therefore, have occurred long before that epoch. The lower region of the ancient valley had reached its mature stage before the rejuvenation commenced, and must have been fashioned during some part of the Tertiary era.

As regards the volume of the river which eroded it, the effect of the rejuvenation has apparently enlarged somewhat the drainage area of the Towy system at the expense of that of neighbouring systems, and a tributary such as the Cothi has undoubtedly suffered considerable changes; these have had the effect of diverting some of the water which flowed into the ancient river near Rhandirmwyn, so that it now enters the Towy many miles lower at Nantgaredig. It is believed, however, that the change in the area of the drainage system as a whole was probably small.

The volume of a river depends on other factors than the drainage area. Among these the rainfall is the most important, and although it is not possible to estimate the rainfall of the Tertiary era, general considerations suggest that it was lower than at present on account of smaller relief. On the other hand the area may have been nearer the sea than at present, a circumstance which might or might not increase the rainfall, or the climate may have been radically different, perhaps tropical or subtropical. A smaller rainfall on the same area as that of the existing drainage system connotes a smaller volume in the river, which would therefore require a higher gradient to enable it to carry the same kind of load as the existing river. Further, with lower relief, the run-off would be less than at present; while if the area were under tropical conditions, an abundant vegetation would also affect the run-off.

As regards the load, the maturity of the ancient drainage system denotes a long preceding period of denudation, during which the rocks of the area were covered by a thick mantle of weathered products. The type of material carried into the ancient rivers would therefore probably consist largely of fine silt and mud, whereas the modern rivers derive their load to a great extent from glacial materials which are on the whole much coarser than the products of rock decay. It may therefore be assumed that the load was of finer grade than at present, and would have required a smaller gradient for its transportation. The influence of a probable smaller rainfall and of a probable finer load are in opposite directions, and, as a first approximation, we may assume their effects to balance.

The second assumption made above is a more serious one, and for its full discussion a much larger region than is dealt with in the present investigation would have to be taken into account. I hope

Longitudinal Profiles of the Upper Towy Drainage System

to return to this problem in due course; for the present, it is assumed that no tilting has occurred. The combined effects of the conditions discussed have left the mature valleys, with minor exceptions, of such a form that the present rivers find them adapted to their needs without serious modification. On the other hand it is possible that, even if a tilting of the old valleys had occurred subsequently to their formation, little change of their form by the present rivers would take place above the head of rejuvenation, so that in the upper reaches of a valley the effect of an uplift accompanied by tilting might be difficult to distinguish from that of a simple uplift.

III. ROCK STEPS AND HANGING VALLEYS

The rock steps and hanging valleys appear to stand in a systematic interrelationship. The most marked drop in the floor of the Towy valley occurs about half a mile above Nant Stalwyn, where the river emerges (4·1) from a narrow gorge. The rock floor in this locality is at least 40 to 50 ft higher above the smoothed long profile than it is in the reach immediately downstream. The valley floor of the tributary Nant Tadarn, if prolonged, would stand 30 to 50 ft above the smoothed profile of the Towy below the gorge, and would thus appear to be at grade with the valley floor above the rock step. Another tributary, Hirnant (4·3), shows an estimated hang of anything between 40 and 100 ft.

Lower downstream, Nant y Gerwyn (5·0) behaves in a similar manner. If no allowance is made for reduction of former gradient at its lower end, this tributary valley would have a hang of 45 ft; but if reduction of gradient downstream is allowed for, its hang could be as much as 120 ft. The valley of the Maesnant stream (5·6) near Nant yr Hwch is estimated to hang about 40 to 80 ft above the Towy, while Nant Cwm-du, which now enters at 6·0 but formerly entered at 6·2, and another tributary, Nant y Fleiddiast, both appear to grade with a level considerably above that of the main valley. On the whole the discordance between the hanging tributary valleys and the floor of the trunk Towy increases downstream.

An abrupt fall in the valley floor, such as occurs above Nant Stalwyn, combined with a series of hanging valleys farther downstream, produces a combination of features which has frequently been attributed to glacial erosion. But the local conditions can be shown not to have been particularly favourable for intense glacial scour; it is more probable that the hanging valleys are, in the main, the result of an uplift which rejuvenated the main valleys and the tributaries alike. On this view the rock step above Nant Stalwyn may be regarded as the rejuvenation head related to the uplift in question.

The valley above the Nant Stalwyn rock step must, then, be the product of a period of base-levelling earlier than that to which the valley between Nant Stalwyn and Fanog corresponds.

Farther down the Towy, again, there is evidence of two distinct periods of base-levelling. The lowest reach of the Trawsnant stream (10·4) is approximately at grade with the present trunk, whereas the middle reach, if prolonged to the Towy, would stand at about 780 ft, i.e. at nearly the same level deduced from the profile of the old Towy floor. The upper reach, however, if similarly prolonged, would stand at about 890 ft. The aspect of the uppermost reach proves that it must have formed part of a drainage system in a late-mature stage of development.

The tributary Nant Lletty-gleison (10·3) on the opposite side of the valley shows similar features. The mature floor is preserved to just below the 1,000-ft contour. If we prolong it, assuming a gradient equal to that between 1,250 and 1,000 ft, it would come into the Towy valley at 820 ft; but if a reasonable allowance is made for downstream diminution of gradient, then the reconstructed former junction falls at the more probable height of about 880 ft.

It is possible to use the values obtained for these and other former tributary junctions in reconstructing an approximate former profile for the floor of the main valley. Extrapolations using the lowest surviving gradients provide minimum values of height, while allowance for reduction of gradient in destroyed portions of the profiles gives values which may be taken as maximal. Although the data cannot supply an accurate guide, a smooth curve drawn below the upper and lower limits thus indicated, between a point above Nant Stalwyn to Bwlchyffin, gives at least an approximate indication of what the long profile used to be.

By means of superposing on this curve either the profile of the existing river or the profile constructed from the logarithmic formula (taking $a = 2·4$), it is possible to obtain an approximate estimate of the base-level to which the restored profile was related. This level appears to be about 580 ft above present sea-level. There is no doubt that detailed levelling of minor tributary valleys would

Longitudinal Profiles of the Upper Towy Drainage System

enable the older valley system to be reconstructed with greater accuracy than is possible at present.

Thus, within the limits of the Towy valley, three distinct base-levels can be distinguished. The lowest or Llandovery base-level corresponds to the existing level of the sea; the middle or Fanog base-level, to which the valley between Nant Stalwyn and Fanog is due, now stands at about 400 ft above Ordnance Datum; while the highest or Nant Stalwyn base-level probably stands nearly 200 ft higher.

The Camddwr

Near the junction with the Towy the Camddwr is graded with that river, but the profile of the middle reach between Capel Soar and Rhydtalog, if prolonged downstream, would join the Towy at a much higher level. When diminishing gradient is allowed for, in the manner previously outlined, the most likely value of height at the former confluence lies in the range 785–815 ft. It is probable, therefore, that the part of the Camddwr valley which lies between Capel Soar and Rhydtalog is related to the shelf in the Towy valley which stands at about 800 ft. Ordnance Datum. This shelf is in turn related to the Fanog base-level. If the profile of the upper region of the Camddwr is prolonged, it would probably stand at a level of about 900 ft in the Towy valley, and this part of the valley is therefore related to the higher region of the Towy, or, in other words, to the Nant Stalwyn base-level.

The Pysgotwr

The profile of the Pysgotwr can be constructed for a length of $2\frac{1}{4}$ miles down to the point of rejuvenation, which is $3\frac{1}{2}$ miles distant from the Towy. The gradient curve constructed from the profile is approximately a rectangular hyperbola. An expression similar to that which was employed for the Towy may therefore be used to represent it.

From the gradient curve, a is found to be $-1 \cdot 6$. Various values of a above and below $-1 \cdot 6$ were tried in the logarithmic expression, and compared with the curve. The results are given in Table 3.3 which also shows the differences between the calculated levels and those obtained from the smoothed profile.

It will be seen that all these values give fairly close agreement with the profile, the average difference ranging from 7 to 10 in. In general, the agreement is closer for the lower part of the curve

TABLE 3.3
Summary of results for the Pysgotwr

x	y obtained from curve	a = -1.2	diff.	a = -1.3	diff.	a = -1.4	diff.	a = -1.5	diff.	a = -1.6	diff.	a = -1.8	diff.
2.1[1]	1096												
2.3	1061	1058.9	-2.1	1058.4	-2.6								
2.5	1029	1028.2	-0.8	1027.6	-1.4	1027	-2.0						
2.7	1002.5	1002.0	-0.5	1001.6	-0.9	1001.1	-1.4	1000.5	-2.0	999.8	-2.7	977	-1.5
2.9	978.5	979.3	0.8	979.0	0.5	978.8	0.3	978.4	-0.1	978.1	-0.4	941.8	0.8
3.3	941	941.1	0.1	941.3	0.3	941.4	0.4	941.5	0.5	941.5	0.5	926.3	1.3
3.5	925	925.1	0.1	925.2	0.2	925.4	0.4	925.6	0.6	925.8	0.8	911.9	0.4
3.7	911.5	910.3	-1.2	910.5	-1.0	910.7	-0.8	910.9	-0.6	911.2	-0.3	898.2	0.2
3.9	898	896.8	-1.2	896.9	-1.1	897.2	-0.8	897.3	-0.7	897.6	-0.4	885.2	0.2
4.1	885	884.3	-0.7	884.4	-0.6	884.5	-0.5	884.6	-0.4	884.8	-0.2		
4.3[1]	872.5												
4.5	861.5	861.9	0.4	861.8	0.3	861.6	0.1	861.4	-0.1	861.2	-0.3	860.7	-0.8
Sum of differences			7.9		8.9		6.7		5.0		5.6		5.2
Average difference			0.79		0.89		0.74		0.62		0.70		0.74
Difference between 3.3 and 4.5			3.7		3.5		3.0		2.9		2.5		3.7
Average			0.62		0.58		0.50		0.48		0.44		0.62
Gradient at the Towy		23.1		26.1		29.3		32.8		36.4		44.4	
Level at the Towy (Towy = 725)		746.3		739.7		730.2		721.9		712.6		691.5	

[1] These values, together with x = 3.2, were used in determining the constants of the expression.

than for the upper. Although it is difficult to decide which value should be adopted in preference to the others, the values of a between $-1\cdot4$ and $-1\cdot6$ give the smallest average differences, both over the whole curve and in the lower part. The level of the Pysgotwr at its junction with the Towy ranges from 691·5 ($a = -1\cdot8$) to 746·3 ($a = -1\cdot2$) ft. Some of these values can be proved inadmissible, if it be assumed that the Pysgotwr has the higher gradient at the junction. Only the values of a for the Pysgotwr which make the gradient of that river greater than 25 ft per mile (a for the Towy being taken as 2·4) need be considered. The value $a = -1\cdot8$ gives too straight and too steep a curve; one of the values between $-1\cdot4$ and $-1\cdot6$ must be chosen. These indicate levels at the confluence of 730·2 and 712·6 ft respectively; the mean of 721·4 is close to the 724 ft determined from the Towy profile. I am inclined to adopt $a = -1\cdot6$ as the most likely value, since it produces a small average deviation from the curve, and since the Pysgotwr before meeting the Towy is joined by the Doethie, a circumstance requiring a lessening of slope below the confluence of these two streams. With $a = -1\cdot6$, this change is provided for, while the gradient of the tributary still remains steeper than that of the trunk Towy. In addition, the various calculations take into account the difference of position between the pre-Glacial and the existing confluences.

I was unable to obtain definite proof of two ancient base-levels in the main Pysgotwr valley or in the Doethie valley, although there is some evidence to suggest a base-level about 70 ft higher than the present valley of the Pysgotwr, where this belongs to the Fanog base-level. A high base-level demonstrable on certain feeder streams agrees with the Nant Stalwyn level of the Towy.

The Cothi

Survey of the Upper Cothi valley was limited to the determination of a few heights on the valley floor, by means of the theodolite. When allowance is made for downstream decrease of gradient, these heights suggest that the former level at the confluence with the Towy was at least 677 ft. There is good agreement with the value of 680 ft calculated for the Towy itself. Although the data in this instance are few and the calculations admittedly crude, the results are wholly consistent with those obtained for other valleys. There is no evidence on the Cothi of the Nant Stalwyn base-level; the whole of its ancient valley seems to have been fashioned in relation to the Fanog level.

The physical features of the Towy valley and its tributaries thus reveal the former history of the drainage system in some detail. The uppermost portions of most of the valleys still retain considerable traces of the form which they attained during the late stages of the Nant Stalwyn period of base-levelling. This period appears to have been a prolonged one in which the valleys reached a late-mature stage of development. It was followed by a lowering of the base-level due to general uplift of the region – whether or not accompanied by tilting cannot at present be determined. Most of the mature valleys were destroyed during the Fanog period of base-levelling to within a few miles of their sources. Towards the close of this period the lower portions of the valleys again attained a mature form. Then followed a second uplift, with consequent lowering of the base-level. A second excavation of the lower parts of the valleys gave rise to their present forms, except in so far as they have been modified by some glacial erosion and deposition. Erosion has already proceeded so far that, for about 40 miles above its mouth, the Towy valley has attained for the third time a mature stage which is related to the existing Llandovery base-level. The points on the Towy and its tributaries where the last remnants of the Nant Stalwyn base-level can be recognised lie on an almost straight line drawn from above Nant Stalwyn to near Bonborfa on the Pysgotwr. This line, if prolonged, would pass beyond the head of the Cothi valley and thus confirms the conclusion previously drawn that the old valley belongs to the Fanog period of base-levelling. It is not improbable, however, that the surface of the plateau near the heads of the valleys is mainly the result of erosion during the oldest or Nant Stalwyn period of base-levelling, and that it has not been seriously modified by subsequent erosion. It may therefore be of high antiquity. No dates can be assigned from any evidence in the upper part of the Towy drainage system to the periods of base-levelling and of uplift, but it is hoped that some clue to their age will be obtained when the lower part of the Towy basin has been examined.

REFERENCES

DE LA NOË, G., and DE MARGERIE, E. (Paris, 1888) *Les Formes du Terrain* 205 pp.
PENCK, A. (Stuttgart, 1894) *Morphologie der Erdoberfläche*.

4 Land Sculpture in the Henry Mountains

G. K. GILBERT

THE Colorado Basin is exceptionally well suited to the study of the origin of topographic forms. It has already occasioned notable contributions to the principles of earth sculpture, and its resources are far from exhausted. The following account, based principally on research in the Henry Mountains, restates certain principles of erosion which have been derived from, or confirmed by, work in the Colorado Plateau country.

EROSIONAL PROCESSES

Waves, glaciers and winds all share in the sculpture and degradation of the land. The present account, however, will be limited to the work of rain and running water. The natural processes of the division and removal of rock make up erosion. They are called disintegration and transportation. Transportation is chiefly performed by running water. Disintegration is naturally divided into two parts. So much of it as is accomplished by running water is called corrasion, and that which is not is called weathering.

Stated in their natural order, the three general divisions of the process of erosion are (1) weathering, (2) transportation and (3) corrasion. The rocks of the general surface of the land are disintegrated by weathering. The material thus loosened is transported by streams to the ocean or other receptacle. In transit it helps to corrade from the channels of the streams other material, which joins with it to be transported.

Weathering
In weathering, the chief agents of disintegration are solution, change of temperature, the beating of rain, gravity and vegetation.

The great solvent of rocks is water, but it receives aid from some other substances of which it becomes the vehicle, chiefly products of the formation and decomposition of vegetable tissues. Some rocks are disintegrated by their complete solution, but most are divided into grains by the solution of a portion; fragmental rocks usually lose only the cement by solution, and are thus reduced to incoherence.

The most rigid rocks are cracked by sudden changes of temperature, and the crevices thus begun are opened by the freezing of the water within them. The coherence of the more porous rocks is impaired and often destroyed by the same expansive force of freezing water. The beating of the rain overcomes the feeble coherence of earths, and assists solution and frost by detaching the particles which they have partially loosened. When the base of a cliff is eroded so as to remove or diminish the support of the upper part, the rock thus deprived of support is broken off in blocks by gravity. The process of which this is a part is called cliff erosion or sapping. Plants often pry apart rocks by the growth of their roots, but their chief aid to erosion is by increasing the solvent power of percolating water.

Transportation

A portion of the water of rains flows over the surface and is quickly gathered into streams. A second portion is absorbed by the earth or rock on which it falls, and after a slow underground circulation reissues in springs. Both transport the products of weathering, the latter carrying dissolved minerals and the former chiefly undissolved fragments. Transportation is also performed by the direct action of gravity. In sapping, the blocks which are detached by gravity are by the same agency carried to the base of the cliff.

Corrasion

In corrasion, the agents of disintegration are solution and mechanical wear. Wherever the two are combined, the superior efficiency of the latter is evident. In all fields of rapid corrasion the part played by solution may be disregarded.

The mechanical wear of streams is performed by the aid of hard mineral fragments which are carried along by the current. The effective force is that of the current; the tools are mud, sand and boulders. The most important of them is sand; it is chiefly by the impact and friction of grains of sand that the rocky beds of streams are disintegrated.

Streams of clear water corrade their beds by solution. Muddy streams act partly by solution, but chiefly by attrition.

Streams transport the combined products of corrasion and

weathering. A part of the debris is carried in solution, and a part mechanically. The finest of the undissolved detritus is held in suspension, the coarsest is rolled along the bottom, and there is a gradation between the two modes. There is a constant comminution of all the material as it moves, and the work of transportation is thereby accelerated. Boulders and pebbles, while they wear the stream bed by pounding and rubbing, are worn still more rapidly themselves. Sand grains are worn and broken by the continued jostling, and their fragments join the suspended mud. Finally the detritus is all more or less dissolved by the water, the finest the most rapidly.

In brief, weathering is performed by solution; by change of temperature, including frost; by rain beating; by gravity; and by vegetation. Transportation is performed chiefly by running water. Corrasion is performed by solution and by mechanical wear.

Corrasion is distinguished from weathering chiefly by including mechanical wear among its agencies, and the importance of the distinction will be apparent when we come to consider how greatly and peculiarly this process is affected by modifying conditions.

CONDITIONS CONTROLLING EROSION

The chief conditions which affect the rapidity of erosion are (1) declivity, (2) the character of the rock and (3) climate.

Rate of erosion and declivity
In general, erosion is most rapid where the slope is steepest, but weathering, transportation and corrasion are affected in different ways and in different degrees.

With increase of slope goes increase in the velocity of running water, and with that goes increase in its power to transport undissolved detritus.

The ability of a stream to corrade by solution is not notably enhanced by great velocity, but its ability to corrade by mechanical wear keeps pace with its ability to transport, or may even increase more rapidly. For not only does the bottom receive more blows in proportion as the quantity of transient detritus increases, but the blows acquire greater force from the accelerated current, and from the greater size of the moving fragments. It is necessary, however, to distinguish the ability to corrade from the rate of corrasion, which will be seen to depend largely on other conditions.

Weathering is not directly influenced by slope, but it is reached indirectly through transportation. Solution and frost, the chief agents of rock decay, are both retarded by the excessive accumulation of disintegrated rock. Frost action ceases altogether at a few feet below the surface, and solution gradually decreases as the zone of its activity descends and circulation becomes more sluggish. The rapid removal of the products of weathering stimulates its action, and especially that portion which depends upon frost. If however, the power of transportation is so great as to remove completely the products of weathering, the work of disintegration is thereby checked; for the soil is a reservoir to catch rain as it reaches the earth and store it up for the work of solution and frost, instead of letting it run off at once unused.

Sapping is directly favoured by great declivity.

Rate of erosion and rock texture

Other things being equal, erosion is most rapid when the eroded rock offers least resistance; but the rocks which are most favourable to one portion of the process of erosion do not necessarily stand in the same relation to the others. Disintegration by solution depends in large part on the solubility of the rocks, but it proceeds most rapidly in fragmental rocks with soluble cement and open texture. Disintegration by frost is most rapid in rocks which absorb much water and are feebly coherent. Disintegration by mechanical wear is most rapid in soft rocks.

Transportation is most favoured by those rocks which yield by disintegration the most finely comminuted debris.

Rate of erosion and climate

The influence of climate upon erosion is less easy to formulate. The direct influences of temperature and rainfall are comparatively simple, but their indirect influence through vegetation is complex, and partly in opposition.

Temperature affects erosion chiefly by its changes. Where the range of temperature includes the freezing point of water, frost contributes its powerful aid to weathering; it is only where changes are great and sudden that rocks are cracked by unequal expansion or contraction.

All the processes of erosion are affected directly by the amount of rainfall, and by its distribution through the year. All are accelerated by its increase and retarded by its diminution. When it is concentrated in one part of the year, transportation and corrasion are accelerated and weathering is retarded.

Weathering is favoured by abundance of moisture. Frost accomplishes most when the rocks are saturated, and solution when there is the freest subterranean circulation. But when the annual rainfall is concentrated into a limited season, a larger share of the water fails to penetrate, and the gain from temporary flooding does not compensate for the checking of all solution by a long dry season.

Transportation is favoured by increasing water supply as greatly as by increasing declivity. When the volume of a stream increases, it becomes at the same time more rapid, and its transporting capacity gains by increased velocity as well as by increased volume. Hence the increase in power of transportation is more than proportional to the increase of volume, and the transportation of a stream which is subject to floods is greater than it would be if its water supply were evenly distributed in time.

The indirect influence of rainfall and temperature, by means of vegetation, has different laws. Vegetation is intimately related to water supply. In proportion as vegetation is profuse, the solvent power of percolating water is increased, and on the other hand the ground is sheltered from the mechanical action of rains and rills. The removal of disintegrated rock is greatly impeded by the conservative power of roots and fallen leaves, and a soil is thus preserved. Transportation is retarded. Weathering by solution is accelerated up to a certain point, but in the end it suffers by the clogging of transportation. The work of frost is nearly stopped as soon as the depth of soil exceeds the limit of frost action. The force of raindrops is expended on foliage. Moreover a deep soil acts as a distributing reservoir for the water of rains, and tends to equalise the flow of streams. Hence the general effect of vegetation is to retard erosion, and since the direct effect of great rainfall is the acceleration of erosion, it follows that its direct and indirect tendencies are in opposite directions.

In arid regions the declivities of which are sufficient to give thorough drainage, the absence of vegetation is accompanied by absence of soil. When a shower falls, nearly all the water runs off from the bare rock and the little that is absorbed is rapidly reduced by evaporation. Solution becomes a slow process for lack of a continuous supply of water, while frost accomplishes its work only when it closely follows the infrequent rain. Thus weathering is retarded. Transportation has its work so concentrated by the quick gathering of showers into floods, as to compensate, in part at least, for the smallness of the total rainfall from which they derive their power.

Hence in regions of small rainfall, surface degradation is usually limited by the slow rate of disintegration, while in regions of great rainfall it is limited by the rate of transportation. There is probably an intermediate condition with moderate rainfall, in which a rate of disintegration greater than that of an arid climate is balanced by a more rapid transportation than consists with a very moist climate, and in which the rate of degradation attains its maximum.

Over nearly the whole of the earth's surface there is a soil, and wherever this exists we know that the conditions are more favourable to weathering than to transportation. Hence it is true in general that the conditions which limit transportation are those which limit the general degradation of the surface.

To understand the manner in which this limit is reached, it is necessary to look at the process by which the work is accomplished.

Transportation and comminution

A stream of water flowing down its bed expends an amount of energy that is measured by the quantity of water and the vertical distance through which it descends. If there were no friction of the water upon its channel, the velocity of the current would continually increase, but if, as is the usual case, there is no increase of velocity, then the whole of the energy is consumed in friction. Friction produces inequalities in the motion of the water, and especially induces subsidiary currents more or less oblique to the general onward movement. Some, with an upward tendency, perform the chief work of transportation. They lift small particles from the bottom and hold them in suspension while they move forward with the general current. The finest particles sink most slowly and are carried farthest before they fall. Larger ones are barely lifted, and are dropped at once. Still larger ones are only half lifted; that is, they are rolled over without quitting the bottom. And finally there is a limit to the power of every current: the largest fragments of its bed are not moved at all.

There is a definite relation between the velocity of a current and the size of the largest boulder it will roll. It has been shown by Hopkins that the weight of the boulder is proportioned to the sixth power of the velocity. It is easily shown also that the weight of a suspended particle is proportioned to the sixth power of the velocity of the upward current that will prevent its sinking. But it must not be inferred that the total load of detritus that a stream will transport bears any such relation to the rapidity of its current. The true inference is that the velocity determines the size limit of the detritus that a stream can move by rolling, or can hold in suspension.

Every particle which a stream lifts and sustains is a draught upon its energy, and the measure of the draught is the weight (weighed in water) of the particle, multiplied by the distance it would sink in still water in the time during which it is suspended. If for the sake of simplicity we suppose the whole load of a stream to be of uniform particles, then the measure of the energy consumed in their transportation is their total weight multiplied by the distance one of them would sink in the time occupied in their transportation. Since fine particles sink more slowly than coarse, the same consumption of energy will convey a greater load of fine than of coarse.

Again, the energy of a clear stream is entirely consumed in the friction of flow, the friction bearing a direct relation to velocity. But if detritus be added to the water, then a portion of its energy is diverted to the transportation of the load, at the expense of the friction of flow and hence at the expense of velocity. As the energy expended in transportation increases, the velocity diminishes. If the detritus be composed of uniform particles, then we may say that as the load increases the velocity diminishes. But the diminishing velocity will finally reach a point at which it can barely transport particles of the given size, and when this point is attained the stream has its maximum load of detritus of the given size. But fine detritus requires less velocity for its transportation than coarse, and will not so soon reduce the current to the limit of its efficiency. A greater percentage of the total energy of the stream can hence be employed by fine detritus than by corase.

The friction of flow is a complex affair. The water in contact with the bottom and walls of the channel develops friction by flowing past them, and that which is farther away by flowing past that which is near. The inequality of motion gives rise to cross-currents and there is a friction of these upon one another. The ratio or coefficient of friction of water against the substance of the bed, the coefficient of friction of water against water (the viscosity of water) and the form of the bed, all conspire to determine the resistance of flow and together make up what may be called the coefficient of the friction of flow. The friction depends on its coefficient and on the velocity.

Thus the capacity of a stream for transportation is enhanced by comminution in two ways. Fine detritus, on the one hand, consumes less energy for the transportation of the same weight, and on the other it can utilise a greater portion of the stream's energy.

It follows, as a corollary, that the velocity of a fully loaded stream depends, other things being equal, on the comminution of the material of the load. When a stream has its maximum load of fine detritus, its velocity will be less than when carrying its maximum load of coarse detritus, and the greater load corresponds to the smaller velocity.

It follows also that a stream which is supplied with heterogeneous debris will select the finest. If the finest is sufficient in quantity, the current will be so checked by it that the coarser cannot be moved. If the finest is not sufficient, the next grade will be taken, and so on.

Transportation and declivity

To consider now the relation of declivity to transportation, we shall assume all other conditions to be constant. Let us suppose that two streams otherwise identical differ in the total amount of fall. Their declivities are proportional to their falls. Since the energy of a stream is measured by the product of its volume and its fall, the relative energies of the two streams are proportional to their falls, and hence proportional to their declivities. The velocities of the two streams, depending, as we have seen above, on the character of the detritus which loads them, are the same, and hence the same amount of energy is consumed by each in the friction of flow. And since the energy which each stream expends in transportation is the residual after deducting what it spends in friction from its total energy, it is evident that the stream with the greater declivity will not merely have the greater energy, but will expend a smaller percentage of it in friction and a greater percentage in transportation. Hence declivity favours transportation in a degree that is greater than its simple ratio.

There are two elements of which no account is taken in the preceding discussion, but which need to be mentioned to prevent misapprehension, although they in no way detract from the conclusions.

The first is the addition which the transported detritus makes to the energy of the stream. A stream of water charged with detritus is at once a compound and an unstable fluid. It has been treated merely as an unstable fluid requiring a constant expenditure of energy to maintain its constitution; but since it is a compound fluid, the energy it develops by its descent is plainly greater than the energy pertaining to the water alone, in the precise ratio of the mass of the mixture to the mass of the simple water.

The second element is the addition which the detritus makes to the friction of flow. The coefficient of friction of the compound stream upon its bottom will always be greater than that of the simple stream of water, and the coefficient of internal friction or the viscosity will be greater than that of pure water, and hence for the same velocity a greater amount of energy will be consumed.

The energy which is consumed in the friction of the detritus on the stream bed accomplishes as part of its work the mechanical corrasion of the bed.

Transportation and quantity of water

A stream's friction of flow depends mainly on the character of the bed, on the area of the surface of contact and on the velocity of the current. When the other elements are constant, the friction varies approximately with the area of contact, which in turn depends on the length and form of the channel and on the discharge. For streams of the same length and same form of cross-section, but differing in size of cross-section, the area of contact varies directly as the square root of the discharge. Hence, other things being equal, the friction of a stream on its bed is proportioned to the square root of the discharge. But, as stated above, the total energy of a stream is directly proportional to the discharge, and the total energy is equal to the energy spent in friction, plus the energy spent in transportation. It follows that if a stream changes its discharge without changing its velocity or other characteristics, the total energy will change at the same rate as the discharge; the energy spent in friction will change at a smaller rate, and the energy remaining for transportation will change at a greater rate.

Hence increase in discharge favours transportation in a degree that is greater than its simple ratio.

It follows as a corollary that the running water which carries the debris of a district loses power by subdivision towards its sources; and that, unless there is a compensating increment of declivity, the tributaries of a river will fail to supply it with the full load it is able to carry.

The obstruction which vegetation opposes to transportation is especially effective in that it is applied at the infinitesimal sources of streams, where the force of the running water is least.

A stream which can transport debris of a given size may be said to be competent to such debris. Since the maximum particles which streams are able to move are proportioned to the sixth powers of their velocities, competence depends on velocity. Velocity, in turn, depends directly on declivity and volume, and inversely on load.

In brief, the capacity of a stream for transportation is greater for fine debris than for coarse. Its capacity for the transportation of a given kind of debris is enlarged in more than simple ratio by increase of declivity; and it is enlarged in more than simple ratio by increase of volume. The competence of a stream for the transport of debris of a given fineness is limited by a corresponding velocity.

The rate of transportation of debris of a given fineness may equal the capacity of the transporting stream, or it may be less. When it is less, it is always from the insufficiency of supply. The supply furnished by weathering is never available unless the degree of fineness of the debris brings it within the competence of the stream at the point of supply.

The chief point of supply is at the very head of the flowing water. Rain, falling on material disintegrated by weathering, begins, after it has saturated the immediate surface, to flow off. But it forms a very thin sheet, its friction is great, its velocity is small, and it is competent to pick up only particles of exceeding fineness. If the material is heterogeneous, it leaves the coarser particles. As the sheet moves on it becomes deeper and soon begins to gather itself into rills. As the deepening and concentration of water progresses, either its capacity increases and the load of fine particles is augmented, or, if fine particles are not in sufficient force, its competence increases, and larger ones are lifted. In either case the load is augmented, and as rill joins rill it steadily grows, until the accumulated water finally passes beyond the zone of disintegrated material.

The particles which the feeble initial currents are not competent to move have to wait either until they are subdivided by weathering, or until the deepening of the channels of the rills so far increases the declivities that the currents acquire the requisite velocity, or until some fiercer storm floods the ground with a deeper sheet of water.

Thus rate of transportation, as well as capacity for transportation, is favoured by fineness of debris, by declivity and by quantity of water. It is opposed chiefly by vegetation, which holds together material loosened by weathering.

When the current of a stream gradually diminishes in its course, for example in approaching the ocean, the capacity for transportation also diminishes; and as soon as the capacity becomes less than the load, precipitation begins, the coarser particles being deposited first.

Corrasion and transportation
Where a stream has all the load of a given degree of comminution which it is capable of carrying, the entire energy of the descending water and load is consumed in the translation of the water and load and there is none applied to corrasion. If it has an excess of load its velocity is thereby diminished so as to lessen its competence and a portion is dropped. If it has less than a full load it is in condition to receive more and it corrades its bottom.

A fully loaded stream is on the verge between corrasion and deposition. As will be explained in another place, it may wear the walls of its channel, but its wear of one wall will be accompanied by an addition to the opposite wall.

The work of transportation may thus monopolise a stream to the exclusion of corrasion, or the two works may be carried forward at the same time.

Corrasion and declivity
The rapidity of mechanical corrasion depends on the hardness, size and number of the transient fragments, on the hardness of the rock bed and on the velocity of the stream. The blows which the moving fragments deal upon the stream bed are hard in proportion as the fragments are large and the current is swift. They are more numerous and harder upon the bottom of the channel than upon the sides because of the constant tendency of the particles to sink in water. Their number is increased up to a certain limit by the increase of the load of the stream; but when the fragments become greatly crowded at the bottom of a stream their force is partially spent among themselves, and the bedrock is in the same degree protected. For this reason, and because increase of load causes retardation of current, it is probable that the maximum work of corrasion is performed when the load is far within the transporting capacity.

The element of velocity is of double importance since it determines not only the speed, but to a great extent the size, of the pestles which grind the rocks. The coefficients upon which it in turn depends, namely declivity and discharge, have the same importance in corrasion that they have in transportation.

Let us suppose that a stream at constant discharge is at some point continuously supplied with as great a load as it is capable of carrying. For so great a distance as its velocity remains constant, it will neither corrade downwards nor deposit, but will leave the grade of its bed unchanged. But if in its progress it reaches a place where a smaller declivity of bed gives a diminished velocity, its capacity for transportation will become less than the load and part of the load will be deposited. Or if in its progress it reaches a place where a greater declivity of bed gives an increased velocity, the capacity for transportation will become greater than the load and there will be corrasion of the bed. In this way a stream which has a supply of debris equal to its capacity tends to build up the gentler slopes of its bed and cut away the steeper. It tends to establish a single, uniform grade.

Let us now suppose that the stream, after having obliterated all the inequalities of the grade of its bed, loses nearly the whole of its load. Its velocity is at once accelerated and vertical corrasion begins through its whole length. Since the stream has the same declivity and consequently the same velocity at all points, its capacity for corrasion is everywhere the same. Its rate of corrasion, however, will depend on the character of its bed, variations in which will produce inequalities of grade. But as soon as there is inequality of grade there is inequality of velocity, and inequality of capacity for corrasion; where hard rocks have produced declivities, the capacity for corrasion will be increased. The differentiation will proceed until the capacity for corrasion is everywhere proportioned to the resistance, and no further – that is, until there is an equilibrium of action.

In general, we may say that a stream tends to equalise its work in all parts of its course. Its power inheres in its fall, and each foot of fall has the same power. When its work is to corrade and the resistance is unequal, it concentrates its energy where the resistance is great by crowding many feet of descent into a small space, and diffuses it where the resistance is small by using but a small fall in a long distance. When its work is to transport, the resistance is constant and the fall is evenly distributed by a uniform grade. When its work includes both transportation and corrasion, as is the usual case, its grades are somewhat unequal. The inequality is greatest when the load is least.

On most streams it is the flood stage which determines the grades of the channel. The load of detritus is usually greatest during the highest floods, and power is conferred so rapidly with increase of discharge that in any event the influence of the stream during its high stage will overpower any influence which may have been exerted at a low stage. That relation of transportation to corrasion which subsists when the water is high will determine the grades of the waterway.

Declivity and quantity of water
The conclusions reached in regard to the relations of corrasion and declivity depend on the assumption that the volume of the stream is the same throughout its whole course, and they consequently apply directly to such portions only of streams as are not increased by tributaries. A simple modification will include the more general case of branching streams.

Let us suppose that two equal confluent streams have identical declivity, and are both fully loaded with detritus of the same kind. If the channel down which they flow after union has also the same declivity, then the joint stream will have a greater velocity than its branches, its capacity for transportation will be more than adequate for the joint load and it will corrade its bottom. By its corrasion it will diminish the declivity of its bed, and consequently its velocity and capacity for transportation, until its capacity is equal to the total capacity of its tributaries. When an equilibrium of action is reached, the declivity of the main stream will be less than the declivities of its branches. This result does not depend on the assumed equality of the branches, nor upon their number. It is equally true that in any river system which is fully supplied with material for transportation and which has attained a condition of equal action, the declivity of the smaller streams is greater than that of the larger.

Let us further suppose that two equal confluent streams are only partially loaded, and are corrading at a common rate a common rock. If the channel down which they flow after union is in the same rock and has the same declivity, then the joint river will have a greater velocity, and will corrade more rapidly than its branches. By its more rapid corrasion it will diminish the declivity of its bed, until as before there is an equilibrium of action, the branch having a greater declivity than the main. This result also is independent of the number and equality of the branches; and it is equally true that in any river system which traverses and corrades rock of equal resistance throughout, and which has reached a condition of equal action, the declivity of the smaller streams is greater than that of the larger.

In general we may say that, other things being equal, declivity bears an inverse relation to quantity of water.

An apparent exception to this law, especially noteworthy in the sculpture of badlands, will be described subsequently.

SCULPTURE

Erosion lays bare rocks which were before covered and concealed, and is thence called denudation. It reduces the surfaces of mountains, plateaux and continents, and is thence called degradation. It carves new forms of land from those which existed before, and is thence called land sculpture. In the following pages it will be considered as land sculpture, and attention will be called to certain principles of erosion which are concerned in the production of topographic forms.

Sculpture and declivity

We have already seen that erosion is favoured by declivity. Where the declivity is great the agents of erosion are powerful; where it is small they are weak; where there is no declivity they are powerless. Moreover it has been shown that their power increases with the declivity in more than simple ratio.

It is evident that if steep slopes are worn more rapidly than gentle, the tendency is to abolish all differences of slope and produce uniformity. The law of uniform slope thus opposes diversity of topography, and if not complemented by other laws would reduce all drainage basins to plains. But in reality it is never free to work out its full results, for it demands a uniformity of conditions which nowhere exists. Only a water sheet of uniform depth, flowing over a surface of homogeneous material, would suffice, and every inequality of water depth or of rock texture produces a corresponding inequality of slope and diversity of form. The reliefs of the landscape exemplify other laws, and the law of uniform slopes is merely the conservative element which limits their results.

Sculpture and structure: the Law of Structure

We have already seen that erosion is influenced by rock character. Erosion is most rapid where the resistance is least, and hence as the soft rocks are worn away the hard are left prominent. The differentiation continues until an equilibrium is reached through the law of declivities. When the ratio of erosive action as dependent on declivities becomes equal to the ratio of resistances as dependent on rock character, there is equality of action. In the structure of the earth's crust hard and soft rocks are grouped with infinite diversity of arrangement. They are in masses of all forms, dimensions and positions, and from these forms are carved an infinite variety of topographic reliefs.

In so far as the law of structure controls sculpture, hard masses stand as eminences and soft are carved in valleys.

The Law of Divides

We have seen that the declivity over which water flows bears an inverse relation to the quantity of water. If we follow a stream from its mouth upwards and pass successively the mouths of its tributaries, we find its volume gradually less and less and its grade steeper and steeper, until finally at its head we reach the steepest grade of all. If we draw the profile of the river on paper, we produce a curve concave upwards and with the greatest curvature at the upper end. The same law applies to every tributary and even to the slopes over which the freshly fallen rain flows in a sheet before it is gathered into rills. The nearer the divide the steeper the slope; the farther away the less the slope.

It is in accordance with this law that mountains are steepest at their crests. The profile of a mountain, if taken along drainage lines, is concave outwards as represented in the Fig. 4.1, and this

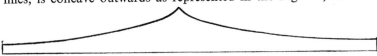

Fig. 4.1 Typical profile of the drainage slope of mountains.

is purely a matter of sculpture, the uplifts from which mountains are carved rarely if ever assuming this form.

Under the Law of Structure and the Law of Divides combined, the features of the earth are carved. Declivities are steep in proportion as their material is hard, and they are steep in proportion as they are near divides. Geological distributions and the distribution of drainage lines and divides are the two sets of conditions on which depends the sculpture of the land. In some areas the first set is the more important, in others the second. In the bed of a stream without tributaries the grade depends on the character of the underlying rock. In homogeneous rock all slopes depend on the distribution of divides and drainage lines. By contrast, in non-homogeneous areas the distribution of drainage lines itself depends in part on geological distributions.

Mountain forms in general depend more on the Law of Divides than on the Law of Structure, but their independence of structure is rarely perfect, and it is difficult to separate the individual results of the two principles.

Again, the relative importance of the two controls is strongly influenced by climate, so much so that climate can rank as a third sculptural control.

Equal action and interdependence

The tendency to equality of action, or to the establishment of a dynamic equilibrium, has already been pointed out in the discussion of the principles of erosion and of sculpture, but one of its most important results has not been noticed.

Of the main conditions which determine the rate of erosion, namely, quantity of running water, vegetation, texture of rock and declivity, only the last is reciprocally determined by rate of erosion. Declivity originates in upheaval or in displacements of the earth's crust, but it receives its distribution in detail in accordance with the laws of erosion. Wherever by reason of change in any of the conditions the erosive agents come to have locally exceptional power, that power is steadily diminished by the reaction of rate of erosion upon declivity. Every slope is a member of a series, receiving the water and the waste of the slope above it, and discharging its own water and waste upon the slope below. If one member of the series is eroded with exceptional rapidity, two things immediately result: first, the member above has its level of discharge lowered, and its rate of erosion is thereby increased; and second, the member below, being clogged by an exceptional load of detritus, has its rate of erosion diminished. The acceleration above and the retardation below diminish the declivity of the member in which the disturbance originated, and as the declivity is reduced the rate of erosion is likewise reduced.

Furthermore, the disturbance which has been transferred from one member of the series to the two which adjoin it is transmitted by them to others, and does not cease until it has reached the confines of the drainage basin. For in each basin all lines of drainage unite in a main line, and a disturbance upon any line is communicated through it to the main line and thence to every tributary. And as any member of the system may influence all the others, so each member is influenced by every other. There is an interdependence throughout the system.

Badlands

The workings of the Law of Divides are best investigated in areas where, since variation of rock texture does not occur, the Law of Structure does not apply. Such areas include badlands.

If we examine a badland ridge, separating two drainage lines and forming a divide between them, we find an arrangement of secondary ridges and secondary drainage lines, similar to that represented in Fig. 4.2.

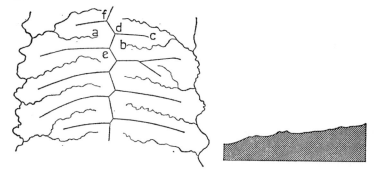

Fig. 4.2 Plan and profile of a badland ridge.

The general course of the main ridge being straight, its course in detail is found to bear a simple relation to the secondary ridges. It follows a zigzag course, being deflected to the right or left by each lateral spur.

The altitude of the main ridge is correspondingly related to the secondary ridges. At every point of union there is a maximum, and in the intervals are saddles. The maxima are not all equal, but bear some relation to the magnitudes of the corresponding secondary ridges, and are especially accented where two or more secondaries join at the same point (see profile in Fig. 4.2).

I conceive that the explanation of these phenomena is as follows: The heads of the secondary drainage lines laid down in the diagram are in nature tolerably definite points. The water which during rain converges at one of these points is there abruptly concentrated in volume. Above the point it is a sheet, or at least is divided into many rills. Below it, it is a single stream with greatly increased power of transportation and corrasion. The principle of equal action gives to the concentrated stream a smaller declivity than to the diffused sheet, and, what is especially important, it tends to produce an equal grade in all directions upwards from the point of convergence. The converging surface becomes hopper- or funnel-shaped; as the point of convergence is lowered by corrasion, the walls of the funnel are eaten back equally in all directions except that of the stream. The influence of the stream in stimulating erosion above its head is thus extended radially and equally through an arc of 180°, of which the centre is at the point of convergence.

Where two streams head near each other, the influence of each tends to pare away the divide between them, and by paring to carry it farther back. The position of the divide is determined by the two influences combined and represents the line of equilibrium between them. The influences being radial from the points of convergence, the line of equilibrium is tangential, and is consequently at right-angles to a line connecting the two points. Thus, for example, if a, b and c (Fig. 4.2) are the points of convergence at the heads of three drainage lines, the divide line ed is at right-angles to a line connecting a and b, and the divides fd and gd are similarly determined. The point d is simultaneously determined by the intersection of the three divide lines.

Furthermore, since that point of the line ed which lies directly between a and b is nearest to those points, it is the point of the divide most subject to the erosive influences which radiate from a and b, and it is consequently degraded lower than the contiguous portions of the divide. The points d and e are less reduced, and d, which can be shown by similar reasoning to stand higher than the adjacent portion of either of the three ridges which there unite, is a local maximum.

There is one other peculiarity of badland forms which is of great significance, but which I shall nevertheless not undertake to explain. According to the Law of Divides, as stated in a previous paragraph, the profile of any slope in badlands should be concave upwards, and the slope should be steepest at the divide. The union or intersection of two slopes on a divide should produce an angle. But in point of fact the slopes do not unite in an angle. They unite in a curve, the profile of a drainage slope, instead of being concave all the way to its summit, changing its curvature and becoming convex. Fig. 4.3 represents a profile from a to b of Fig. 4.2. From a to m

5 Erosional Development of Streams: Quantitative Physiographic Factors

R. E. HORTON

IN spite of the general renaissance of science in the present century, physiography still remains largely qualitative. Streams and their drainage basins are described as youthful, mature, old, poorly drained or well drained, without specific information as to how, how much or why. An effort will be made in the present paper to show how the problem of erosional morphology may be approached quantitatively. Drainage-basin development by ground-water erosion, highly important as it is, will, however, not be considered, and the discussion of drainage development by surface run-off will mainly be confined to processes occurring outside of stream channels. Nor will the equally important phase of the subject, channel development, be considered in detail.

Stream orders
In continental Europe attempts have been made to classify stream systems on the basis of branching or bifurcation. In this system of stream orders, the largest, most branched, main or stem stream is usually designated as of order 1 and smaller tributary streams of increasingly higher orders (Gravelius, 1914). The smallest unbranched fingertip tributaries are given the highest order.

Feeling that the main or stem stream should be of the highest order, and that unbranched fingertip tributaries should always be designated by the same ordinal, the author has used a system of stream orders which is the inverse of the European system. In this system unbranched fingertip tributaries are always designated as of order 1, tributaries or streams of the second order receive branches or tributaries of the first order, but these only; a third-order stream must receive one or more tributaries of the second order but may also receive first-order tributaries. A fourth-order stream receives branches of the third and usually also of lower orders, and so on. In this system the order of the main stream is the highest (Fig. 5.1).

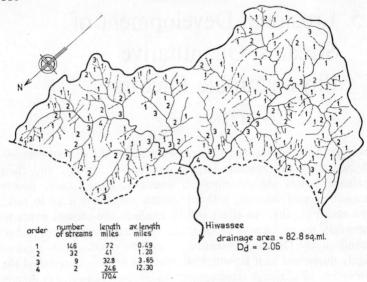

Fig. 5.1 Drainage net, upper Hiwassee river.

Drainage density

Fig. 5.2 shows two small drainage basins, both on the same scale, one well drained, the other poorly drained. These terms, 'well drained' and 'poorly drained', while in common use, are purely qualitative, and something better is needed; the simplest and most convenient tool is drainage density or average length of streams within the basin per unit of area (Horton, 1932). Expressed as an equation:

$$\text{Drainage density, } D_d = \frac{\Sigma L}{A} \quad (5.1)$$

where ΣL is the total length of streams and A is the area, both in units of the same system.[1] The well-drained basin has a drainage density 2·74, the poorly drained one 0·73, or one-fourth as great.

For accuracy, drainage density must, if measured directly from maps, be determined from maps on a sufficiently large scale to show all permanent natural stream channels. Perennial streams and intermittent streams should alike be included. If only perennial streams were included, a drainage basin containing only intermittent streams would, in accordance with equation (5.1), have zero drainage density. Most of the work of valley and stream development by

[1] A consolidated list of symbols is given below, in the Appendix (p. 163).

running water is performed during floods. Intermittent and ephemeral streams carry flood waters, and hence should be included.

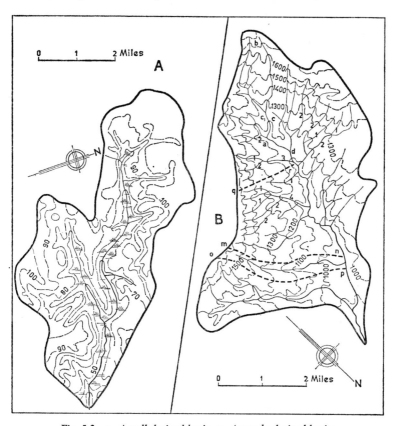

Fig. 5.2 A, *A well-drained basin.* B, *A poorly-drained basin.*

In textbooks on physiography, differences of drainage density are commonly attributed to differences of rainfall or relief, and these differences in drainage density are largely used to characterise physiographic age in the sense used by Davis (Davis, 1909; Wooldridge and Morgan, 1937). In the poorly drained area (Fig. 5.2) the mean annual rainfall is about 30 per cent greater than in the well-drained area. Therefore some other factor or factors must be far more important than either rainfall or relief in determining drainage density. The other factors are infiltration capacity of the soil or terrain and initial resistivity of the terrain to erosion.

Length of overland flow

The term *length of overland flow*, designated l_o, is used to describe the length of flow of water over the ground before it becomes concentrated in definitive stream channels. To a large degree, length of overland flow is synonomous with length of sheet flow as quite commonly used. Overland flow is sustained by a relatively thin layer of surface detention. Channel flow is sustained by accumulated channel storage.

In addition to its obvious value in various ways in characterising the degree of development of a drainage net within a basin, drainage density is particularly useful because of the fact that the average length of overland flow l_o is in most cases approximately half the average distance between the stream channels, and hence is approximately equal to half the reciprocal of the drainage density, or

$$l_o = \frac{1}{2D_d}. \tag{5.2}$$

Later it will be shown that length of overland flow is one of the most important independent variables affecting both the hydrologic and physiographic development of drainage basins.

In this paper it is frequently assumed for purposes of convenience that the average length of overland flow is sensibly equal to the reciprocal of twice the drainage density. The author has shown elsewhere (Horton, 1932) that the average length of overland flow is given by the equation

$$l_o = \frac{1}{2D_d \sqrt{[1 - (s_c/s_g)^2]}} \tag{5.3}$$

where s_c is the channel or stream slope and s_g the average ground slope in the area.

Values of the correction factor or of the ratio $l_o/2D_d$ for different values of the slope ratio s_c/s_g are as follows:

$s_c/s_g =$	0·9	0·8	0·7	0·6	0·5	0·4	0·3	0·2	0·1
$l_o/2D_d =$	1·86	1·67	1·40	1·25	1·15	1·09	1·05	1·02	1·005

The ground slope or resultant slope of the area tributary to a stream on either side is necessarily always greater than the channel slope. Table 5.1 shows the average channel and ground slopes of streams in the Delaware river and some other drainage basins. The channel slope is commonly from one-half to one-fourth of the ground slope. If the channel slope is less than one-third of the

TABLE 5.1
Characteristics of certain drainage nets

Stream	Location	Type	Order of main stream σ	Area (sq. miles) A	No. of streams Σ No.	No. of first-order streams N_1	Stream frequency F_s	Drainage density D_d	Average length of first-order streams L_1	Bifurcation ratio r_b	Length ratio r_l	ΣL
(1)	(2)	(3)	(4)	(5)	(6)	(7)	(8)	(9)	(10)	(11)	(12)	(13)
Esopus Creek	Olive Bridge, N.Y.	Mountains	5	234·0	126	90	0·527	0·849	0·99	3·12	2·31	203·2
Esopus Creek	Lower area[1]	Rolling and plains	7	426·3	361	256	0·847	0·818	0·81	2·27	1·84	348·6
Rondout Creek	Honk Falls, N.Y.	Mountains	4	105·0	58	44	0·552	1·07	1·08	3·30	2·64	112·6
Putnam Brook	Weedsport, N.Y.	Glacial, drumlin	4	27·0	26	18	0·963	1·95	0·77	2·46	2·74	52·7
Cold Spring Brook	Weedsport, N.Y.	Glacial, drumlin	4	15·8	25	15	1·58	2·025	0·58	2·62	2·66	32·0
Crane Creek	Weedsport, N.Y.	Glacial, drumlin	5	45·7	48	31	1·05	2·03	0·81	2·22	2·30	92·6
Ganargua Creek	Lyons, N.Y.	Glacial, drumlin	6	299·0	269	166	0·899	1·628	0·87	2·89	2·30	487·1
Keuka Lake	Foot of Lake, N.Y.	Hilly, dissected	5	161·0[3]	170	124	1·055	1·665	1·16	3·25	1·96	268·0
Seneca Lake	Foot of Lake, N.Y.	Hilly, dissected	6	479·0[3]	472	334	0·984	1·59	0·95	3·15	2·20	762·5
Owasco Lake	Weedsport, N.Y.	Hilly, dissected	5	200·0	265	191	1·325	1·79	0·83	3·91	2·22	358·0
Thunder Bay River	Alpena, Mich.[2]	Glacial, flat	4	—	44	33	—	—	—	3·00	—	—

[1] Ashakan Dam to Saugerties, N.Y.   [2] Data furnished by Prof. C. O. Wisler.   [3] Land area, excluding lake.

ground slope, the error resulting from the assumption that average length of overland flow is equal to the reciprocal of twice the drainage density may in general be neglected.

Stream frequency

This is the number of streams, F_s, per unit of area, or

$$F_s = \frac{N_s}{A} \tag{5.4}$$

where N_s = total number of streams in a drainage basin of A areal units.

Values of drainage density and stream frequency for small and large drainage basins are not directly comparable, because they usually vary with the size of the drainage area. A large basin may contain as many small or fingertip tributaries per unit of area as a small drainage basin, and in addition it usually contains a larger stream or streams. This effect may be masked by the increase of drainage density and stream frequency on the generally steeper slopes of smaller drainage basins.

COMPOSITION OF DRAINAGE NET

The term *drainage pattern* often implies little more than the manner of distribution of a given set of tributary streams within a drainage basin. Thus, with identically the same lengths and numbers of streams, the drainage pattern may be dendritic, rectangular or radial. Neither the drainage pattern nor the drainage density adequately characterises the stream system or drainage net. Something more is needed as a basis for quantitative morphology of drainage basins. The author has therefore coined the expression *composition of a drainage net*, as distinguished from *drainage pattern*. Composition implies the numbers and lengths of streams and tributaries of different sizes or orders, regardless of their pattern. Composition has a high degree of hydrologic significance, whereas pattern alone has but little hydrologic significance, although it is highly significant in relation to geology.

Cotton (1935) and others have used the term *texture* to express composition of a drainage net as related both to drainage density and stream frequency. For quantitative purposes two terms are needed, since two drainage nets with the same drainage densities may have quite different numbers and lengths of streams. Numerical values of drainage density independent of other units are needed for various purposes.

LAWS OF DRAINAGE COMPOSITION

The numbers and lengths of tributaries of different orders were determined for the streams listed in Table 5.1. The numbers and the lengths of streams varied with the stream order in a manner which suggested a geometrical progression. On plotting the data on semilogarithmic paper, it was found (Fig. 5.3) that the stream numbers

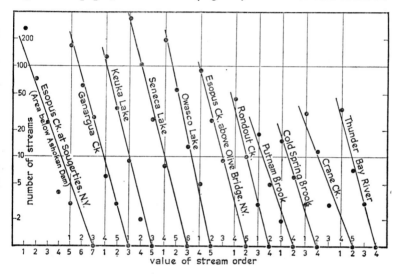

Fig. 5.3 Bifurcation, or relation of stream order to stream number.

fall close to straight lines, and that the same is true of the stream lengths (Fig. 5.4). From the manner of plotting, these lines are necessarily graphs of geometrical series, inverse for stream numbers of different orders and direct for stream lengths.

From the properties of geometric series, the equation of the lines giving the number N_o of streams of a given order in a drainage basin can be written

$$N_o = r_b^{(s-o)} \tag{5.5}$$

From the laws governing geometric series it is easily shown that the number N of streams of all orders is

$$N = \frac{r_b^s - 1}{r_b - 1} \tag{5.6}$$

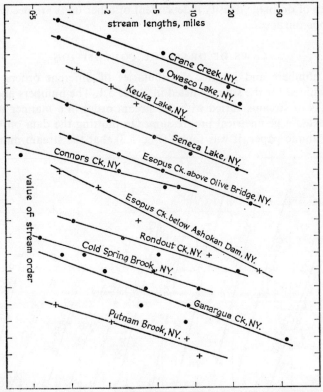

Fig. 5.4 Stream length and stream order.

By definition, o is the order of a given class of tributaries, s is the order of the main stream and r_b is the bifurcation ratio.

The equation correlating the lengths of streams of different orders is, similarly,

$$l_o = l_1 r_l^{o-1} \tag{5.7}$$

As an example, Table 5.2 shows the observed and computed numbers and lengths of streams of different orders, based on the following values of the variables:

$$r_b = 3\cdot 12$$
$$s = 5$$
$$r_l = 2\cdot 31$$
$$l_1 = 0\cdot 994$$

The data given in Table 5.1 cover a wide range of conditions, from precipitous mountain areas, like upper Esopus Creek, and highly dissected areas, like those of Seneca and Owasco lakes, to moderately rolling and flat areas. They also cover drainage basins

TABLE 5.2

Observed and computed stream lengths and stream numbers, drainage basin of Esopus Creek above Olive Bridge, N.Y.

Stream order	Number of streams		Average stream length (miles)	
	From topographic maps	By equation (5.5)	From topographic maps	By equation (5.7)
1	90	94·75	0·994	0·994
2	25	30·37	2·45	2·30
3	9	9·73	5·64	5·31
4	1	3·12	6·00	12·2
5	1	1·00	29·00	28·3

ranging in size from a few square miles up to several hundred square miles.

The bifurcation ratio (Table 5.1, col. 11) ranges from about 2 for flat or rolling drainage basins up to 3 or 4 for mountainous or highly dissected drainage basins. As would be expected, the bifurcation ratio is generally high for hilly, well-dissected drainage basins. The values of the length ratios (col. 12) range from about 2 to about 3; the average is 2·32.

If stream lengths in this table had been measured as extended to watershed lines, the resulting stream-length ratios would have been materially reduced. If c is the average length from the stream tip to the watershed line, and l_1 and l_2 are actual average stream lengths of two successive orders, l_2 being the higher, then, as computed in Table 5.1,

$$r_l = \frac{l_2}{l_1} \qquad (5.8)$$

If measured as extended to the watershed lines,

$$r'_l = \frac{l_2 + c}{l_1 + c} \qquad (5.9)$$

The quantity r'_l will always be less than r_l. The average value of r_l for the streams listed in Table 5.1 is 2·32. If the stream lengths were extended to the watershed lines, this value would lie between 2·00 and 2·32. The theoretical value of r'_l for streams flowing into larger streams at right-angles is 2·00, but r'_l will be greater for streams entering at acute angles, as do most streams on steep slopes. The distance along the course of a stream from its mouth extended to the watershed line is called *mesh length*. The use of this quantity instead of actual stream length is preferable in physiographic studies; its use leads to closer agreement with the theoretical values.

In Figs. 5.3 and 5.4 the agreement between the mean lines for the different streams and the observed data is so close that the two following general laws may be stated regarding the composition of stream-drainage nets:

1. *Law of Stream Numbers.* The numbers of streams of different orders in a given drainage basin tend closely to approximate an inverse geometric series in which the first term is unity and the ratio is the bifurcation ratio.

2. *Law of Stream Lengths.* The average lengths of streams of each of the different orders in a drainage basin tend closely to approximate a direct geometric series in which the first term is the average length of streams of the first order.

Total length of streams of a given order

Since the total length of streams of a given order is the product of the average length and number of streams, equations (5.5) and (5.7) can be combined into an equation for total stream length of a given order.

The total length L_o of tributaries of order o is:

$$L_o = l_1 r_b^{s-o} r_l^{o-3} \tag{5.10}$$

The total lengths of all streams of a given order is the product of the number of streams and length per stream. The number of streams is dependent on the bifurcation ratio r_b and increases with stream order, while the length per stream is dependent on the stream length r_l and decreases with increasing stream order. Thus the total lengths of streams of a given order should have either a maximum or a minimum value for some particular stream order. A maximum or minimum may not occur, because the stream order required to give the maximum or minimum stream length may exceed the order of the main stream, in which case the total lengths of streams of a given order will either increase or decrease progressively with increasing stream order. An exception occurs where r_b and r_l are equal. Then the total lengths of streams of all orders are the same and equal to $l_1 r_b^s$. The ratio of r_l to r_b is designated ρ and is an important factor in relation both to drainage composition and to physiographic development of drainage basins. As will be shown later, the value of the ratio $\rho = r_l/r_b$ is determined by precisely those factors, hydrologic, physiographic, cultural and geological, which determine the ultimate degree of drainage development in a given drainage basin.

By summation of the total stream lengths for different orders, as given by equation (5.10), the total stream length within a drainage basin can be expressed in terms of four fundamental quantities: l_1, o_s, r_b and r_l.

General Equation of Composition of Stream Systems
From equation (5.10) the total length of streams of a given order is:
$$L_o = l_1 r_b^{s-o} r_l^{o-1}.$$

The total length of all streams in a drainage basin with the main stream of a given order s is the sum of the total lengths of streams of different orders, or:

$$\Sigma L = l_1 [r_b^{s-1} + r_b^{s-2} r_l + r_b^{s-2} r_l^2 + \ldots r_b^o r_b^{s-1}] \quad (5.11)$$

This equation is, however, cumbersome. It can readily be shown that drainage density is given by:

$$D_d = \left(\frac{l_1 r_b^{s-1}}{A}\right)\left(\frac{\rho^s - 1}{\rho - 1}\right) \quad (5.12)$$

This equation combines all the physiographic factors which determine the composition of the drainage net of a stream system in one expression. Aside from its scientific interest in this respect, it can also be used to determine drainage density. Values of the factor $(\rho^s - 1)/(\rho - 1)$ can be obtained from Fig. 5.5.

Relation of size of drainage area to stream order
Since equation (5.12) incorporates all the principal characteristics of the stream system of a drainage basin, it may be considered a quantitative generalisation of Playfair's law. It can be written in such a form as to give any one of the quantities l_1, D_d, A, r_b, r_l and s when the other five quantities are known. If the ratio $\rho < 1$, then, for larger values of s, ρ^s is small, and $\rho^s - 1$ may be taken as $-1 \cdot 0$. The equation (5.12) may then be written:

$$s = 1 + \frac{\log[(1-\rho)D_d A/l_1]}{\log r_b} \quad (5.13)$$

E

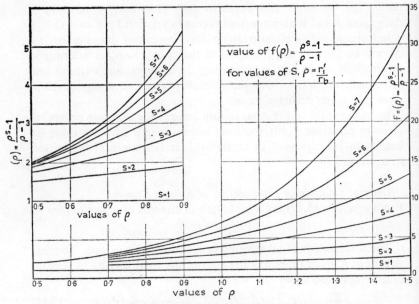

Fig. 5.5 Diagram for $\rho^s - 1/\rho - 1$.

If $\rho > 1$, then, for large values of s, ρ^s is sensibly the same as $\rho^s - 1$, and $\rho^s - 1$ is positive. This leads similarly to the equation:

$$(s-1)\log r_b + s\log\rho = \log\frac{(\rho-1)D_d A}{l_1} \qquad (5.14\text{A})$$

and

$$s = \frac{\log[(\rho-1)D_d A/l_1] + \log r_b}{\log r_b + \log\rho} \qquad (5.14\text{B})$$

The quantity $\log r_b$ is small relative to $\log(\rho-1)D_d A$ and may be neglected, so that in either case the order of the main stream developed in a drainage basin of a given area A increases for larger values of s in proportion to the logarithm of the area A. If, for example, with given values of ρ, D_d, r_b, r_l and l_1, an area of 10,000 sq. miles is required to develop a stream order s, then, under the same conditions, in a drainage basin of 100,000 sq. miles, the main stream would be of order one unit higher, and in a drainage basin of 1 million sq. miles the main stream would be two units higher in order than in an area of 10,000 sq. miles. This shows at once why stream systems with extremely high orders do not occur; there is not room for the requisite drainage basins on the surface of the earth.

Law of stream slopes

The relation of stream slope to stream order in a given drainage basin is hinted in Playfair's law. As an illustration, the slopes of streams orders in the Neshaminy, Tohickon and Perkiomen drainage

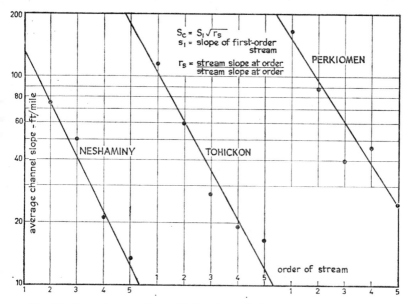

Fig. 5.6 Law of stream slopes: Neshaminy, Tohickon and Perkiomen basins.

basins have been plotted in terms of stream order (Fig. 5.6); there is a fairly definite relationship between slope and order.

Determination of physiographic factors for drainage basins

To determine completely the composition of a stream system, it is necessary to know: (1) the drainage area, A; (2) the order s of the main stream; (3) the bifurcation ratio r_b; (4) the stream-length ratio r_l; and (5) the length l_s of the main stream or preferably the average length l_1 of first-order streams. From these data the drainage density, stream frequency and other characteristics of the stream system can be determined by means of the equations which have been given.

From equation (5.5), $N_o = r_b^{(s-o)}$; then, if $o = s - 1$,

$$N_{s-1} = r_b \qquad (5.15\text{A})$$

This shows that the bifurcation ratio r_b is equal to the number of streams of the next to the highest order for the given drainage basin.

If the stream numbers for different stream orders are plotted on semi-log paper, (Fig. 5.3), the bifurcation ratio r_b can be determined by simply reading from the average line the number of streams of the second-highest order.

From equation (5.7), $l_o = l_1 r_1^{o-1}$; then, if $o = 2$,

$$\frac{l_2}{l_1} = r \tag{5.15B}$$

The stream-length ratio r_l can therefore be obtained by dividing the average stream length of any order by the average length stream of the next lower order, the values of stream lengths being read from the diagram of stream lengths plotted in terms of stream order. It is preferable to use these data rather than actual measured values, since actual numbers can only be integers. Fig. 5.1 illustrates stream numbering, with first-order streams marked by dotted lines.

The determination of stream lengths and orders by direct measurement from maps which are on a sufficiently large scale to show all first-order streams is so laborious as to be practically prohibitive except for smaller drainage basins.

Fortunately, all the required quantities – l_s, l_1, r_b, r_l, D_d – can be determined from smaller-scale maps from which the lower-order tributaries are omitted. The maps must show correctly the streams for several of the higher orders. The order of the main stream is of course unknown, since it is not in general known which of the lower orders of streams are omitted from the map. The streams shown are assigned orders assuming that the main stream has an unknown order s, the next lower order of stream shown is designated 2, and so on. The number of streams of each assumed order is counted, their stream lengths measured from the map, the results tabulated as follows and plotted as shown in Fig. 5.7, line A.

Data for Perkiomen Creek

Order	Assumed inverse order	Number of streams	Average stream length (miles)
s	1	1	20·5
$s-1$	2	2	13·75
$s-2$	3	10	3·61
$s-3$	4	32	1·39

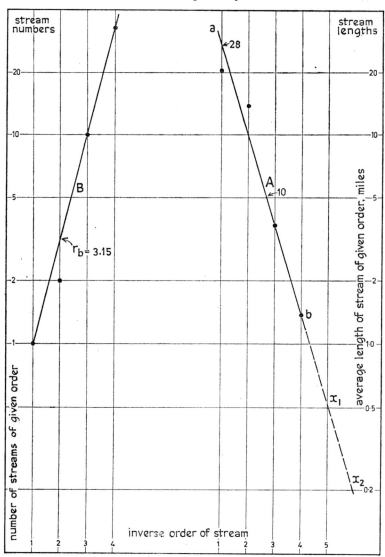

Fig. 5.7 Graphical determination of stream characteristics.

If it is assumed that the main stream is of the

4th order, then $l_1 = 1 \cdot 38$ miles;
5th order, then $l_1 = 0 \cdot 50$ mile;
6th order, then $l_1 = 0 \cdot 20$ mile.

Since l_1 is not far from half a mile, the main stream is of the fifth-order. From line B (Fig. 5.7) the number of second-order streams is 3·15. This is the bifurcation ratio. From line A the lengths of second- and first-order streams are, respectively, 1·38 and 0·52 miles. This gives the stream-length ratio:

$$r_l = \frac{1·38}{0·52} = 2·70.$$

Data for at least four stream orders are required to determine the order of the main stream from incomplete data by this method. Care must also be used in determining the lines A and B accurately to secure correct results.

The values of the stream lengths as far as known are then plotted on semi-log paper (Fig. 5.7, line A) in terms of inverse stream orders, a line of best fit is drawn to represent the plotted points, and this line is extended downwards to stream length unity or less.

To determine the order of the main stream it is necessary to know the order of magnitude but not the exact value of the average length of streams of the first order. It is rarely less than a third of a mile greater than 2 or 3 miles. The point at which the stream length shown by the line ab (Fig. 5.7, line A) extended downwards has a value about the same as the known value of l_1 for the given order indicates the order of the main stream.

This method for determining the order of the main stream is of limited value, particularly in large basins which are not homogeneous, but for small and reasonably homogeneous basins it is accurate. Proof of its validity is readily obtained by applying this method to a drainage basin where the values of l_1 and the drainage density D_d have been determined from measurements on a map showing streams of all orders, but using in the determination only the data for streams of higher orders. This was done in preparing Fig. 5.7, which is of the fifth order, although only data for the first four stream orders were used in the computation, it being assumed that l_1 was of an order of magnitude between 1 and 1·5.

This determination of s also gives the average length l_1 of first-order streams. The bifurcation ratio r_b and the stream-length ratio r_l are determined by the slopes of the lines A and B on Fig. 5.7. It is not necessary to know the order of the main stream to determine these quantities. When r_b, r_l, A, s and l_1 are known, the drainage density can be determined by means of equation (5.12). This method of determining s has the advantage that it is at least as accurate when applied to large as when applied to smaller drainage basins.

Table 5.3 shows the drainage composition of the Neshaminy, Tohickon and Perkiomen Creek stream systems, derived in the manner described, together with the drainage densities as computed by equation (5.12) and as derived from direct measurement from topographic maps.

TABLE 5.3

Observed and computed drainage densities, Neshaminy, Tohickon and Perkiomen Creeks

	Stream	Neshaminy	Tohickon	Perkiomen
1	Stream	Neshaminy	Tohickon	Perkiomen
2	Location	(below forks)	(Point Pleasant)	(near Frederick)
3	Drainage area (sq. miles)	139·3	102·2	152·0
	Computed values:			
4	Stream order from map	5	5	5
5	l_s	35·0	33·0	27·0
6	r_b	3·45	3·00	3·15
7	r_l	2·92	2·85	2·70
8	$\rho = r_l/r_b$	0·85	0·95	0·86
9	$f(\rho)$	3·75	4·50	3·80
10	r_b^{s-1}	141·6	81·0	98·4
11	r_b^{s-1}/A	1·02	0·79	0·65
12	(11) × (9)	3·82	3·56	2·47
13	l_1	0·50	0·53	0·52
14	(12) × (13) = D_d	1·91	1·89	1·28
15	Drainage density from map	1·60	1·91	1·24

Drainage densities computed by equation (5.12) will usually be somewhat higher than those derived directly from maps if stream lengths are measured directly and only to the fingertips of the stream channels, because the stream lengths and mesh lengths are sensibly identical for higher-order streams, whereas there may be 10 to 25 per cent or even 50 per cent difference between stream length and mesh length for low-order streams. In computing drainage density from values of l_1, r_b and r obtained graphically, the computed value corresponds more nearly to drainage density expressed in terms of mesh length than in terms of actual stream length for lower-order streams.

Relation of geological structures to drainage composition

The examples of drainage composition shown in Table 5.1 and in Figs. 5.3 and 5.4 in nearly all cases represent special or abnormal conditions, such as the presence of large lakes in several of the drainage basins, but the table nevertheless agrees well with the laws of stream numbers and stream lengths. Fig. 5.8 shows two drainage basins, the boundaries of which are fixed by geological

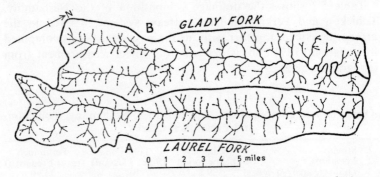

Fig. 5.8 Drainage patterns of Laurel and Glady Forks, Cheat river basin.

structures. Data of stream lengths and stream numbers in these basins are as in Table 5.4.

TABLE 5.4

Data of stream lengths and stream numbers for Fig. 5.8

	Order	Length (miles)	Number of streams	Average length (miles)
Basin A–Laurel Fork 50·8 sq. miles	1	47·80	94	0·51
	2	23·36	26	0·90
	3	4·10	4	1·02
	4	19·62	1	19·62
		94·88		
$D_d = \dfrac{94\cdot 88}{50\cdot 8} = 1\cdot 87$				
Basin B–Glady Fork 55·44 sq. miles	1	38·42	95	0·41
	2	34·76	37	0·94
	3	6·50	6	1·08
	4	20·80	1	20·80
		100·48		
$D_d = \dfrac{100\cdot 48}{55\cdot 44} = 1\cdot 82$				

Erosional Development of Streams

In both streams the law of stream numbers is closely obeyed. The law of stream lengths is approximately obeyed for lower-order streams, but in both basins it is necessary, in order that the area should be drained, that the main stream should have a length

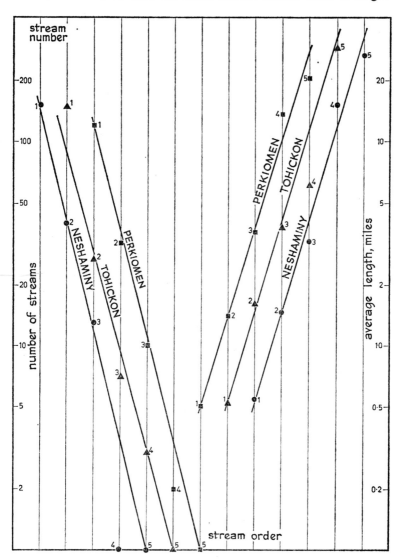

Fig. 5.9 Stream numbers and stream lengths: basins as in Fig. 5.6.

sensibly equal to that of the drainage basin; this requires that the length of the main stream should be much greater than it ordinarily would be for a drainage basin of the same order, of normal form.

One may naturally ask whether stream systems in similar terrain and which are genetically similar should not have identical or nearly identical stream composition. Data for the drainage basins of Neshaminy, Tohickon and Perkiomen creeks in the Delaware river drainage basin near Philadelphia, Pennsylvania, are given in Table 5.3 and in Fig. 5.9. The drainage patterns of these streams are shown in Fig. 5.10. The physiographic characteristics of these three drainage basins are closely similar.

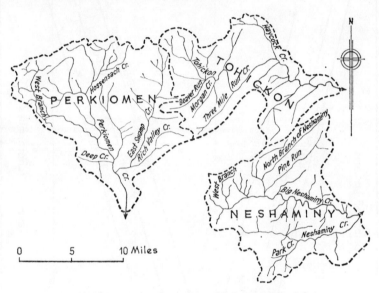

Fig. 5.10   Drainage patterns: basins as in Fig. 5.6.

Tables 5.5 and 5.6 show, respectively, drainage composition of streams in the upper Delaware river drainage basin and drainage composition of several small tributaries of Genesee river in western New York. The streams in the upper Delaware river drainage basin are generally similar, with the exception of Neversink river, in topography, geology and climate. The various morphological factors for these basins are of the same order of magnitude, although not numerically identical. The tributaries of Genesee river listed in Table 5.6 represent areas at various locations around the margins of this

TABLE 5.5
Physiographic factors for drainage basins tributary to the Delaware river

Item		Formula or Method	East Branch Delaware	Beaver Kill	Beaver Kill	Little Delaware	East Branch Delaware	Willowemoc	Beaver Kill	Neversink
(1)		(2)	(3)	(4)		(5)	(6)	(7)	(8)	(9)
Stream			E. Br. Del. R.	Beaver Kill	Beaver Kill	Little Del. R.	E. Br. Del. R.	Willowemoc	Beaver Kill	Neversink
Location			Fish Eddy	Cooks Falls		Delhi	Downsville	Liv'ton M.	Beaver Kill	Claryville
Drainage area (sq. miles)		A	785	241		50·3	373	63·8	61·0	68·6
Length of main stream (miles)		L_s	56·5	28·5		15·0	38·0	14·5	13·0	13·0
Width ratio		A/L_s^2	0·246	0·297		0·224	0·258	0·303	0·361	0·406
Channel slope (ft. per mile)	S	$\dfrac{\Sigma E_1 - \Sigma E_2}{\Sigma L}$	338	210		296	406	224	307	336
Ground slope (ft. per mile)	S_g		1000	779		959	1074	725	1033	1043
Slope ratio	r_s	S_c/S_g	0·338	0·270		0·308	0·377	0·309	0·297	0·323
Order of main stream	(a)	s	5	4		4	5	4	3	3
Average length of first-order streams	(a)	l_1	0·93	0·72		0·82	0·88	0·80	1·05	0·87
Bifurcation ratio	(a)	r_b	4·2	5·0		3·0	3·6	2·7	4·5	4·6
Stream-length ratio	(a)	r_l	2·77	3·12		2·59	2·60	2·54	3·51	3·71
Ratio p	(a)	r_l/r_b	0·66	0·62		0·86	0·72	0·94	0·78	0·80
Average length of overland flow miles		$l_o = \dfrac{1}{2D_d}\sqrt{1-(s_c/s_g)^2}$	0·346	0·405		0·324	0·317	0·360	0·401	0·438
Drainage density		D_d	1·36	1·19		1·47	1·46	1·32	1·19	1·08
Latitude (N.)			42·00	42·00		42·15	42·10	42·00	42·05	41·55
Longitude (W.)			74·50	74·40		74·50	74·40	74·40	74·40	74·30

[1] From Bien's Atlas of New York State. Other data are from U.S. Geological Survey maps.

TABLE 5.6

Physiographic factors for tributaries of the Genesee river, western New York State

Item		Slader Creek	Gates Creek	Rush Creek	Red Creek	Spring Creek	Stony Creek
(2)		(3)	(4)	(5)	(6)	(7)	(6)
Stream		Slader Crk.	Gates Crk.	Rush Crk.	Red Crk.	Spring Crk.	Stony Crk.
Location		Canaseraga	at mouth	Fillmore	at mouth	at mouth	at mouth
Drainage area (sq. miles)		15.8	20.2	43.3	25.3	20.3	22.2
Width ratio		1.12	1.26	0.39	0.65	1.44	0.89
Order of main stream	s	3	4	3	3	3	4
Stream numbers							
First order		24	32	38	14	15	25
Second order		4	10	10	4	4	8
Third order		1	3	1	1	1	3
Fourth order			1				1
Average stream length (miles)							
First order		0.61	0.49	0.48	1.29	1.22	0.81
Second order		2.12	0.95	1.15	1.50	2.00	1.28
Third order		3.75	2.75	10.50	6.25	3.75	3.25
Fourth order			4.00				5.00
Average stream slope (ft. per mile)							
First order		271	89	171	26	37	287
Second order		121	41	80	18	39	196
Third order		89	39	49	7	11	169
Fourth order			25				120
Bifurcation ratio	r_b	4.40	3.10	8.10	3.90	3.90	3.10
Stream-length ratio	r_l	2.21	2.00	4.59	2.57	1.78	1.85
Total stream length	mi	27.00	37.50	60.25	30.25	30.00	38.75
Drainage density	D_d	1.71	1.86	1.41	1.20	1.47	1.73
Ratio ρ	n/r_b	0.50	0.64	0.57	0.66	0.46	0.60
Latitude (N.)		42.25	42.50	42.25	43.05	43.05	42.30
Longitude (W.)		77.45	77.30	78.05	77.40	78.05	77.40

basin between Lake Ontario and the New York–Pennsylvania state line, and comprise a wider range of geological and topographical conditions than occurs in the Delaware river drainage basins, there are correspondingly greater variations in the morphological factors, particularly bifurcation ratio, length of first-order streams and drainage density.

It is found from plotting stream lengths and stream orders, subject to the limiting conditions already described, that both these laws are quite closely followed. Departures from the two laws will, however, be observed, and if other conditions are normal these departures may in general be ascribed to effects of geological controls. As a rule the law of stream numbers is more closely followed than the law of stream lengths; Nature develops successive orders of streams by bifurcation quite generally in a uniform manner, regardless of geological controls. Stream lengths, on the other hand, may be definitely limited by geological controls, such as fixed boundaries of the outline of the drainage basin.

INFILTRATION THEORY OF SURFACE RUN-OFF

The infiltration theory of surface run-off (Horton, 1935, 1937, 1938a) is based on two fundamental concepts:
1. There is a maximum limiting rate, the infiltration capacity (Horton, 1933), at which the soil when in a given condition can absorb rain as it falls.
2. When run-off takes place from any soil surface, large or small, there is a definite functional relation between the depth of surface detention δ_a or the quantity of water which accumulates on the soil surface, and the rate of surface run-off or channel inflow q_s.

Infiltration capacity

The infiltration capacity f of a given terrain, including soil and cover, is controlled chiefly by (1) soil texture, (2) soil structure, (3) vegetational cover, (4) biological structures in the soil and moisture content of the soil, and (5) condition of the soil surface (Horton, 1940; Duley and Kelly, 1941). Temperature is probably also a factor, although its effect is often masked by biological factors, which also vary with temperature and season.

The infiltration capacity of a given area is not usually constant during rain, but starting with an initial value f_o, it decreases rapidly at first, then after about half an hour to two or three hours attains a constant value f_c. The relation of the infiltration capacity to duration of rain can be expressed by the following equation, with f, f_o and f_c in inches per hour:

$$f = f_c + (f_o - f_c)e^{-\kappa_f t} \tag{5.16}$$

where e is the base of Napierian logarithms, t is time from beginning of rain, in hours, and κ_f is a proportionality factor (Horton, 1939, 1940).

As an example typical of many experimental determinations of the change of infiltration capacity during rain, the values of f have been computed at different times, t, from the beginning of rain for a soil with initial infiltration capacity $f_o = 2 \cdot 14$ in. per hour which drops to a constant value $f_c = 0 \cdot 26$ in. per hour in two hours. The quantity κ_f determines the rate of change of infiltration capacity during rain for a given rain intensity and in this case has the value 3·70.

t	0·0	0·2	0·4	0·6	0·8	1·0	1·5	2·0
f (in. per hour)	2·14	1·16	0·69	0·46	0·36	0·31	0·28	0·26

These values of f show that the infiltration capacity drops off rapidly at first, then more slowly as it approaches f_c. Between rains, drying out of the soil and restoration of crumb structure leads to restoration of f towards or to its initial value. The geomorphic significance of this decrease of infiltration capacity during rain is illustrated by the fact that if infiltration capacity remained constant at the high value it usually has at the beginning of rain, there would be little surface run-off or soil erosion. It is the minimum value f_c of infiltration capacity which predominates during most of long or heavy rains, which are chiefly effective in producing floods and sheet erosion. *Transmission capacity* is the volume of flow per unit of time through a column of soil of unit cross-section, with a hydraulic gradient unity or with a hydraulic head equal to the length of the soil column. Infiltration capacity and transmission capacity are related but not identical; the infiltration capacity is usually less than the transmission capacity.

Rain falling at an intensity i which is less than f will be absorbed by the soil surface as fast as it falls and will produce no surface run-off. The rate of infiltration is then less than the infiltration capacity and should not be designated as infiltration capacity. If the rain intensity i is greater than the infiltration capacity f, rain

Erosional Development of Streams

will be absorbed at the capacity rate f; the remaining rain is called *rainfall excess*. This produces run-off. The difference between rain intensity and infiltration capacity in such a case is denoted by σ and designated the *supply rate* ($\sigma = i - f$). For a constant rain intensity i, in inches per hour, the run-off intensity q_s, in inches per hour, approaches the supply rate σ asympototically as a maximum or limiting value as the rain duration increases (Horton, 1939; Beutner et al., 1940). The total surface run-off is approximately equal to the total supply σt_e, where t_e is the duration of rainfall excess.

Overland or sheet flow

The use of sheet flow to describe overland flow not concentrated in channels larger than rills is appropriate, but may not imply flow to depths measured in feet or even in inches but rather in fractions of an inch.

Since 1 in. per hour equals approximately 1 sec.-ft per acre or 640 c.s.m.,[1] and an acre is 208 ft square, the surface run-off intensity q_1 in cubic feet per second from a unit strip 1 ft wide and a slope length l_o will be:

$$q_1 = 0.000023 l_o q_s \quad (5.17)$$

where q_s is the run-off intensity in inches per hour. Discharge = depth × velocity, or, if δ is the depth of sheet flow, in inches, and v the velocity in feet per second, $q_1 = v\delta/12$. It follows that the depth of sheet flow at any point on a slope where the slope length is l_o will be:

$$\delta = \frac{0.000277 l_o q_s}{v} \quad (5.18)$$

On a gently sloping lawn, with a length of overland flow of 100 ft and a velocity of 0.25 ft per second, a depth of surface detention of 0.11 in. will produce 1 in. run-off per hour.

Law of overland flow

The velocity of turbulent hydraulic flow is expressed in terms of the Manning formula:

$$v = \frac{1.486}{n} R^{2/3} \sqrt{S} \quad (5.19)$$

where v is the mean velocity in feet per second, n is the roughness factor, having the same general meaning for sheet flow as for

[1] c.s.m. = cubic feet per second per square mile.

channel flow, R is the hydraulic radius or ratio of area of cross-section to wetted perimeter. For sheet or overland flow, R becomes identical with the depth δ. S is the slope, expressed hydraulically as the ratio fall/horizontal length. For steeper slopes, the sine of the slope angle should be used in place of S.

The run-off intensity in inches per hour from a strip of unit width, for turbulent flow, can be expressed by:

$$q_s = \kappa_s \delta^{5/3}{}_c \tag{5.20A}$$

where K_s is a constant for a given strip of unit width, having a given slope, roughness and slope length.

A similar equation:

$$q_s = \kappa_l \delta^z S \tag{5.20B}$$

can be derived from Poiseuille's law for non-turbulent or laminar flow.

Overland flow may be either wholly turbulent, wholly laminar or partly turbulent and partly laminar. Since the equations for turbulent and laminar flow are of the same form, it follows that the relation between depth of surface detention and run-off intensity, in inches per hour, should be a simple power function of the depth of surface detention, or:

$$q_s = \kappa_s \delta^M \tag{5.20C}$$

where q_s is the run-off intensity in inches per hour, δ is the depth of surface detention at the lower end of the slope, in inches, κ_s is a coefficient involving slope, length of overland flow, surface roughness and character of flow, and the exponent M has a value of 5/3 for fully turbulent flow.

Except for very slight depths of surface detention, this simple law of surface run-off is remarkably well verified by plot experiments (Fig. 5.11). The circles indicate points derived directly from the hydrograph, and the solid lines the resulting relation curves plotted logarithmically. The points indicate an accurate functional relationship between δ_a and q_s.

Except on steep slopes there are always depressions, often small but numerous, on a natural soil surface. If the derived points were plotted for smaller depths than those shown in Fig. 5.11, the corresponding relation lines would curve off to the left, indicating that the power-function relation of q_s to detention depth changes for very slight depths of surface detention. This represents the effect of depression storage.

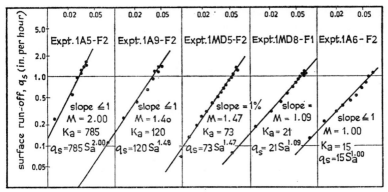

Fig. 5.11 Relation of surface run-off intensity, q_s (in. per hour) to average depth (in.) of surface detention: Concho Basin experiment plots.

Profile of overland flow

The profile of sheet or overland flow, or the relation of depth δ of surface detention to the distance x downslope from the watershed line, is expressed by a simple parabolic or power function (Horton, 1938a). This relation is illustrated in Fig. 5.12. A similar

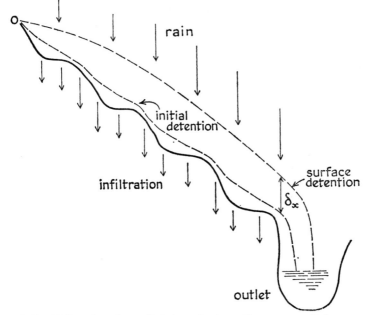

Fig. 5.12 Half-section of a small drainage basin, to illustrate run-off phenomena.

power function expresses the relation of velocity of overland flow in terms of distance from the watershed line (Horton, 1937). For turbulent flow:

$$\delta_x = \left(\frac{\sigma x}{\kappa_s l_o}\right)^{3/5} \tag{5.21A}$$

$$v_x = \frac{1 \cdot 486}{n}\left(\frac{\delta_2}{12}\right)^{2/3} \sqrt{S} \tag{5.21B}$$

and in general, for δ_x in inches, for any type of flow:

$$\delta_x = M\sqrt{\frac{\sigma x}{\kappa_s l_o}} \tag{5.21C}$$

and for velocity in feet per second:

$$v = 0 \cdot 2836 \frac{\sqrt{S}}{n}\left(\frac{\sigma x}{\kappa_s l_o}\right)^{\frac{2}{3m}} \tag{5.21D}$$

where σ = supply rate, in inches per hour; l_o = total length of slope on which overland flow occurs, in feet; x = distance, in feet, downslope from the watershed line; κ_s is a coefficient derived from the Manning formula for turbulent flow, and approximately applicable to other types of flow. Its value is:

$$\kappa_s = \frac{1020\sqrt{S}}{In l_o} \tag{5.22A}$$

and the exponent:

$$\frac{2}{3M} = \frac{2}{9 \cdot 0 - 4I} \tag{5.22B}$$

in which S is the slope, I the index of turbulence, l_o the length of overland flow, and n is a roughness factor, of the same type as the roughness factor in the Manning formula. The equations for turbulent flow are derived directly from the Manning formula. The equations for other types of flow are closely approximate. The depth δ_x as given by these equations is the total depth of surface detention, including depression storage. In the equations both for turbulent and for other types of flow, it is assumed that the velocity varies as $\sqrt{S}$, as for turbulent flow. This may not be entirely correct, although numerous experiments indicate that, in mixed flow, most of the resistance is that due to turbulence. Theoretically, for laminar flow the velocity should vary directly as the slope, not as $\sqrt{S}$. These equations apply primarily to steady flow. Experiments show that they are, however, closely approximate during the early stages of run-off, while surface detention is building up to its maximum value.

The equations for depth and velocity profiles, in conjunction with that for K_s, are of fundamental importance in relation to erosional conditions, since they express the two factors, δ_a and v_a, which control the eroding and transporting power of sheet flow, in terms of the independent variables which govern surface run-off phenomena. There are six variables: (1) rain intensity, i; (2) infiltration capacity, f; (3) length of overland flow, l_o; (4) slope, s; (5) surface roughness factor, n; (6) index of turbulence or type of overland flow, I. To apply these equations to erosion and gradational problems one must also have laws governing the relation of velocity and depth of overland flow to the eroding and transporting power of overland flow.

SURFACE EROSION BY OVERLAND FLOW

Soil-erosion processes

There are always two and sometimes three distinct but closely related processes involved in surface erosion of the soil: (1) tearing loose of soil material; (2) transport or removal of the eroded material by sheet flow; (3) deposition of the material in transport or sedimentation. If (3) does not occur, the eroded material will be carried into a stream.

The spots most vulnerable to erosion are the steeper portions of the hill or valley slopes, neither at the crest nor at the bottom of the hill but intermediate. All soils possess a certain resistivity to erosion, and this resistivity may be increased greatly by a vegetation cover, especially a good grass sod. The underlying soil may have a much smaller resistivity to erosion, and if the surface conditions are changed by cultivation or otherwise so as to destroy the surface resistance, erosion will begin on land which has not hitherto been subject to it.

Fig. 5.13 shows a half-profile of a typical stream valley slope, with the vertical scale greatly exaggerated. The line $oabc$ represents the soil-surface profile, flat in the region o, near the crest, steepest in the region ab, about mid-length of the slope, and relatively flat at the foot of the slope, in the region bc. The line $odef$ represents the surface of sheet or overland flow in an intense rain, the depth of overland flow increasing downslope from o towards f. In the region oa no erosion occurs throughout a distance x_c from the crest of the slope, and this is called the belt of no erosion. Here the energy of the sheet or overland flow is not sufficient to overcome the initial

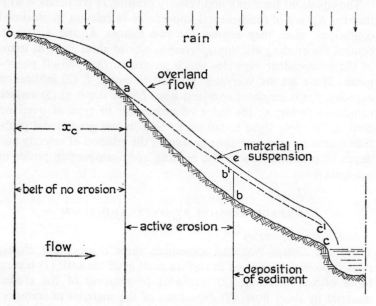

Fig. 5.13 Half-profile of valley slope, to illustrate soil-erosion processes.

resistance of the soil surface to erosion, even in the most intense storm. In the belt *ab*, mid-length of the slope and where the slope is steepest, active erosion occurs.

Eroding force

Erosion by aqueous agencies involves three processes: (1) dislodgement or tearing loose of soil material and setting it in motion (this is called *entrainment*); (2) transport of material by fluid motion; (3) sedimentation or deposition of the transported material.

Let x = distance from divide or watershed line, measured on and along the slope (not horizontally); δ_x = depth of overland flow at x, in inches; w_1 = weight per cubic foot of water in run-off, including solids in suspension; α = slope angle; v = velocity of overland flow at x, in feet per second. The energy expended in frictional resistance per foot of slope length on a strip 1 ft wide running down the slope, per unit of time, for steady flow, will be equal to the energy of the volume of water passing over a unit of area per unit of time. This

Erosional Development of Streams

is the product of the weight, fall and velocity, or:

$$e = w_1 \frac{\delta_x}{12} v \sin \alpha \tag{5.23A}$$

The energy equals the force times the distance moved. Hence the force exerted parallel with the soil surface per unit of slope length and width is:

$$F_1 = \frac{e}{v} = w_1 \frac{\delta_x}{12} \sin \alpha \tag{5.23B}$$

Equation (5.23B), known as the DuBoys formula, is a rational expression which may be used as a basis for determining the eroding force per square foot of soil surface. F_1 is the force available to dislodge or tear loose soil material.

As shown in connection with surface run-off, for turbulent flow the depth of overland flow at the distance x from the watershed line can be expressed in terms of slope, run-off intensity and surface roughness:

$$\delta_x = \left(\frac{\sigma n x}{1020}\right)^{3/5} \left(\frac{1}{s^{0.3}}\right) \tag{5.24A}$$

The slope s is ordinarily expressed as the tangent of the slope angle α or as $\tan \alpha$. Also, for steady overland flow, $\sigma = q_s$, in inches per hour. Substituting these values in (5.24A):

$$d_x = \left(\frac{q_s n x}{1020}\right)^{3/5} \left(\frac{1}{s^{0.3}}\right) \tag{5.24B}$$

Sudstituting this value of δ_x in (5.23B) gives as the total eroding force at x:

$$F_1 = \frac{w_1}{12}\left(\frac{q_s n x}{1020}\right)^{3/5} \left(\frac{\sin \alpha}{\tan^{0.3} \alpha}\right) \tag{5.24C}$$

Critical length x_c: belt of no erosion
Erosion will not occur on a slope unless the available eroding force exceeds the resistance R_i of the soil to erosion. The eroding force increases downslope from the watershed line (equation (5.24C)). The distance from the watershed line to the point at which the eroding force becomes equal to the resistance R is called the *critical distance* and is designated x_c. Between this point and the watershed line no erosion occurs. This strip adjacent to the watershed line, and immune to erosion, is designated the *belt of no erosion*. An expression for the width of the belt of no erosion can readily be

obtained from equation (5.24c) by substituting R for F_1, making $x = x_c$, and solving the equation for x_c. The run-off is free from sediment where erosion begins, and $w_1 = 62.4$ lb. per cu. ft.

$$1020 \left(\frac{12}{62.4}\right)^{5/3} = 65.0 \tag{5.24d}$$

Substituting this constant in (5.24c) gives:

$$x_c = \frac{65}{q_s{}^n} \left(\frac{R_i}{f(s)}\right)^{5/3} \tag{5.24e}$$

where $f(s)$ is a function expressing the effect of slope on the critical length x_c, and is given by the equation:

$$f(s) = \frac{\sin \alpha}{\tan^{0.3} \alpha} \tag{5.24f}$$

For slopes less than 20°, $f(s)$ increases nearly in proportion to the slope. The critical length x_c varies inversely as the run-off intensity q_s in inches per hour, inversely as the roughness factor n, and directly as the 5/3 power of the resistance R_i (equation (5.24e)).

Table 5.7 gives numerical values of x_c for $R_i = 0.01$, 0.05, 0.10, 0.20 and 0.50 lb. per sq. ft, for slope angles of 5°, 10° and 20°, and for four different run-off intensities. These are computed for the roughness factor $n = 0.10$, but can easily be applied to other roughness factors, since the value of x_c is the same if the product $q_s n$ is the same.

The critical length x_c is the most important factor in relation to the physiographic development of drainage basins by erosion processes and also in relation to erosion control. The value of x_c (Table 5.7) is highly sensitive to changes in the variables by which it is controlled, in particular the resistance R_i and the run-off intensity q_s. With a newly cultivated bare soil, with R_i small, 0.05 lb. per sq. ft, for example, with a 10° slope and a run-off intensity of 1 in. per hour, the width of the belt of no erosion would be 35.1 ft, whereas on the same terrain, but with a good, well-developed grass sod to protect the soil, and R_i increased to 0.5 lb. per sq. ft, the belt of no erosion would be 1,573 ft wide. The width of the belt of no erosion varies with the rain intensity.

Most slopes do not have a uniform gradient from the watershed line to a stream. The belt of no erosion will usually comprise all the upper, flatter portion of the slope. If, for example, the slope length is 2,000 ft, $q_s = 1.5$ in. per hour, and the mid-portion of the slope has a gradient of 10° and a resistivity of 0·5 lb. per sq. ft, erosion will begin 1,049 ft from the watershed line. If the lower 250 ft of the slope is flatter (its gradient being 5°), then the length of overland flow required to produce erosion with this slope would be 2,350 ft.

TABLE 5.7

Critical length x_c for various values of R_i

Slope angle α, and run-off intensities q_s, with roughness factor

$$n=0\cdot10: x_c=\frac{65}{q_s n}\left(\frac{R_i}{(fS)}\right)5/3$$

R_i	α (degrees)	q_s (in. per hour)			
		0·5	1·0	1·5	2·0
		$1/q_s n$			
		20	10	6·67	5·00
0·01	5	10·66	5·33	3·35	2·67
0·01	10	4·80	2·40	1·60	1·20
0·01	20	2·28	1·14	0·76	0·57
0·05	5	153·4	76·7	51·16	38·3
0·05	10	70·2	35·1	23·41	17·6
0·05	20	32·6	16·3	10·87	8·2
0·10	5	487·6	243·8	162·61	121·9
0·10	10	218·4	109·2	72·84	54·6
0·10	20	102·8	51·4	34·28	25·7
0·20	5	1535·2	767·6	512·0	383·8
0·20	10	689·0	344·5	229·8	172·2
0·20	20	325·0	162·5	108·4	81·3
0·50	5	7046·0	3523·0	2349·8	1762·0
0·50	10	3146·0	1573·0	1049·2	787·0
0·50	20	1478·0	739·0	4929·0	369·0

Consequently no erosion would occur on the lower or flatter portion of the slope. This example shows why erosion is generally confined to the steeper, middle portion of a given slope (Fig. 5.13).

Rain intensity and erosion

Maximum rain intensities in a given locality generally occur in storms of the summer thunderstorm type. Some soils are easily pulverised when excessively dry, but possess coherence through the operation of capillary force when partially dried after a gradual wetting. If an abrupt intense rain occurs on such a non-coherent soil, the soil may be beaten into a pasty semi-fluid mass by rain of high intensity before run-off begins and before the soil surface becomes protected by surface detention. Such a semi-fluid mass of soil may be carried into the stream by surface run-off *en masse*.

The combination of the conditions described frequently produces what is referred to as a *cloudburst flood*. The term 'cloudburst flood' is used because of the characteristics of the flood rather than those of the rain which produces it. Measurements show that, while such floods may carry large volumes of solids, they often carry surprisingly little water as run-off.

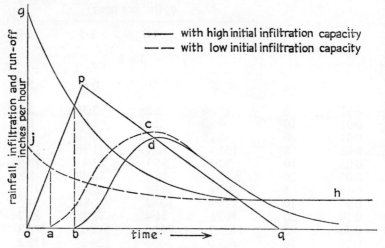

Fig. 5.14   Relation of infiltration capacity to erosion.

In Fig. 5.14, with a rain pattern *opq*, conditions for initially high and initially low infiltration capacity are shown by solid and by dotted lines respectively. With a high initial infiltration capacity *og* there is no surface detention or run-off during the interval *ob* during which rain intensity has risen nearly to the maximum. With a previously wet and packed soil and low initial infiltration capacity *oj*, surface detention and run-off begin earlier at *a*, while the rain intensity is still low. These conditions produce a wall of turbid water or fluid mixed with debris travelling down the stream channel.

The solid material carried along in cloudburst floods is popularly believed to be derived from stream banks and channels, but most of it may in fact be derived from the upland. When the flood wave debouches from a mountain canyon, the water in and behind it may escape laterally or by infiltration. The mud flow then slows down and finally stops, and an accumulation of mud flows may form a debris cone (Horton, 1938b; Bailey, 1935; Bailey et al., 1934).

Hydraulic conditions do not permit the occurrence of shallow steady flow on steep slopes (Horton, 1938b; Jeffreys, 1925); the run-off water is concentrated in a succession or train of more or less uniformly spaced waves. These waves concentrate the impact of overland flow on irregularities or obstructions on the surface and greatly accentuate surface erosion. While grass or close-growing crops are in general an excellent preventive of erosion, they may contribute to active upland erosion in cloudburst floods. As long as the grass remains standing it decreases the velocity but increases the depth of surface detention, and the resistance of run-off is exerted at right-angles to the grass stems and has little tendency to pull the sod loose (Fig. 5.15). When a certain depth and velocity of overland flow is attained the grass begins to flatten down, like the tipping down of a row of dominoes standing on end by pushing the first domino against its neighbour. Then (Fig. 5.15B) the pull of frictional resistance is exerted parallel with the surface. A dense grass sod may be torn up and rolled down the slope like a snowball (Fig. 5.15C). The less resistant underlying soil is thus exposed to erosion.

Transportation and sedimentation

The transportation of eroded material by overland flow or in stream channels takes place in various ways: (1) as bed load; (2) as suspended load; (3) as material held permanently in suspension by molecular agitation (the Brownian movement); (4) in chemical solution. This last mode of transport, while it is the most important process in connection with ground-water erosion, is relatively low in order of importance in connection with surface run-off. As the terms are ordinarily applied in connection with the dynamics of streams, there is no sharp line of demarcation between bed load and suspended load, the former term applying to material carried along, on, or near the solid boundary. Flat fragments are transported by sliding or as bed load. Round particles may be transported by a combination of rolling, sliding and jumping. The process of transport of particles by hopping from point to point in semi-elliptic arcs has been described (Gilbert, 1914) as saltation.

In turbulent flow, eddies thrown off at the solid boundary surfaces have an upward component of velocity. At the same time there is a gradual settling of the water between eddies. The upward velocity and the magnitude and frequency of eddies increase with the velocity of flow and with the roughness of the boundary surface.

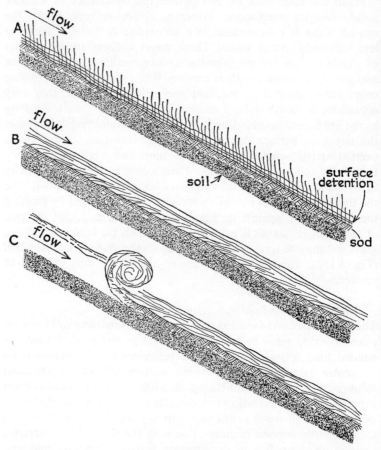

Fig. 5.15 Erosion of sodded area initiated by the breaking-down of grass cover in intense rains.

Saltation in its simplest form involves the picking-up of solid particles by ascending eddies. Those lifted and transported by the field are usually carried only a short distance; those entrapped near the centre of the section of a vortex ring may be carried much farther, until they are thrown out of the ring by centrifugal force.

These two processes are more or less distinct, although both depend on the laws of vortex motion. For this reason mathematical analyses of bed load and suspension transport without taking vortex motion into account are likely to prove inadequate and unsatisfactory.

Most of the work done on sediment transport has been in connection with stream channels. Turbulent flow consists of laminar flow on which is superposed the effect of the transverse motion of eddies. If the flow is turbulent, then only a minute fraction of the energy consumed would be required to provide an equal mean velocity of laminar flow. The remaining energy becomes, in effect, latent at the boundary by conversion into rotational energy of vortex motion.

Two principal results follow: (1) the mean velocity is reduced from that for laminar to that for turbulent flow; (2) the velocity distribution is changed from that for laminar to that for turbulent flow.

For the usual slight depths of sheet flow the energy actually used in translational motion of the fluid is a much larger fraction of the total energy available than for types of flow commonly occurring in stream channels. The relative roughness is usually much greater for overland flow than for channel flow. Sand particles 0·001 ft in diameter with overland flow 0·01 ft in depth correspond in relative roughness to boulders 1 ft in diameter in a stream channel 10 ft in depth. Because of these and other differences, the extent to which experiments and analyses for channels are applicable to sediment transport in sheet or overland flow is an open question.

Much more work is needed on this problem. However, the following facts appear to be well established:

1. The transporting power of sheet flow increases with the amount of eddy energy due to surface resistance.
2. Kinetic energy varies as the square of velocity, and transporting power of sheet flow must vary at least as the square, and perhaps as some higher power, of the velocity.
3. There is a maximum or limiting volume of eroded material which can be transported in suspension by a unit volume of overland flow at a given velocity.

ORIGIN AND DEVELOPMENT OF STREAM SYSTEMS AND THEIR VALLEYS BY AQUEOUS EROSION

Rill channels and rilled surface

The first step toward the gradation of newly exposed sloping terrain is the development of shallow parallel gullies wherever the length of overland flow is greater than the limiting critical distance x_c. These are *rill channels*; a surface covered with such channels is a *rilled surface*. Rill channels are usually relatively uniform, closely spaced and nearly parallel channels of small dimensions which are initially developed by sheet erosion on a uniform, sloping, homogeneous surface.

On some newly exposed lands, with high infiltration capacity and high resistivity to erosion, the length of slope from the major divide to the downslope edge of the area may never exceed x_c. Under these conditions a rilled surface may not develop. This condition often occurs on sand-dune areas and in some glaciated areas with deep permeable soils, especially where grass or other vegetation cover develops soon after the disappearance of glacial ice. In the latter case, x_c on the newly formed surface may exceed the values of l_o pertaining to the drainage of meltwater from the ice sheet; a rilled surface may not develop, and the topography will remain much the same as when the ice disappeared, except that gradation by solution may take place. In desert regions, with suitable relations between the rain intensity, infiltration capacity, surface resistivity and the slope, a rilled surface may develop with little or no crossgrading, so that surface gradation may never extend beyond the rill stage.

Origin of rill channels

Surface run-off starts at the watershed line as true sheet flow, without channels. Even below the critical distance x_c it should apparently continue as such sheet flow combined with sheet erosion. But channels start to develop where there is an accidental concentration of sheet flow. They often form on new terrain without vegetation cover and with a value of R_i sensibly the same at and to some depth below the soil surface. Slight accidental variations of topography may produce a sag in which the depth of sheet flow is a maximum at the point a (Fig. 5.16), the line bb' representing the water surface at maximum run-off intensity. As a result of the greater depth at *a*, erosion will be most rapid at that point, and increased channel capacity will be provided at *a*; part of the water which originally flowed in shallower depths on the adjacent area will be diverted into this enlarged channel. This may accelerate the

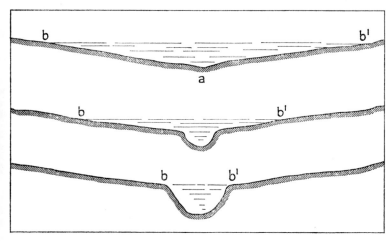

Fig. 5.16 Succession of rill-channel development.

process until the entire flow is concentrated in the rill channel (Fig. 5.16). When a rill channel has once formed, sheet flow coming down the slope upstream from the head of the rill will be deflected towards and diverted into the rill channel, providing a means of rapid headward extension of the rill.

The ultimate dimensions of a stream channel are, as indicated by Playfair's law, such that it is adapted to the area which it drains. Stream channels tend to acquire ultimate dimensions such as to carry all or most of the flood waters of the stream. This is largely because most surface erosion and channel erosion occurs during floods.

Cross-grading and micropiracy
A system of parallel rills is transformed to a dendritic drainage net as the result of the tendency of the water to flow along the resultant slope lines, and is a direct consequence of the overtopping and breaking down of intermediate ridges between gullies by overland flow during heavier storms.

The deepest and widest rill develops where the net length $l_o - x_c$ in which erosion can occur is greatest. If x_c varies, this may not occur where the total length l_o of overland flow is greatest. The longest, deepest and strongest rill channel will be called the master rill. Owing to smaller values of $l_o - x_c$, proceeding away from the master rill on each side, the rills will be shallower, or, considering two adjacent rills, the bottom of the one farther from the master rill will be higher.

When a storm occurs exceeding in intensity preceding storms on the newly exposed areas, the divide between two rills may be broken down at its weakest point by (1) caving-in of the divide; (2) erosion by the deeper or lower rill, diverting the higher rill; (3) overtopping of the divide at the low point by the higher rill, again diverting it into the lower rill. This breaking down of divides between adjacent rill channels is described as *micropiracy*. Micropiracy obliterates the original system of rills and their intermediate ridges on a uniform newly exposed surface. The process of erosion in a stream system and its accompanying valleys destroys most of the record of their origin. Ultimately the original slope parallel with the stream is replaced on each side of the stream by a new slope deflected towards the stream. This process is described as *crossgrading*.

DRAINAGE-BASIN TOPOGRAPHY

Marginal belt of no erosion: gradation of divides

In addition to controlling the drainage density and the composition of the drainage pattern and fixing the end-point of development of a stream system on a given area, the critical distance x_c and the belt of no erosion which it produces govern the degree of gradation which can occur on a given area, and the extent of gradation along and adjacent to both exterior and interior watershed lines or divides.

If the angle between the watershed line aa' (Fig. 5.17A) and the direction of overland flow is A, then for a given critical length x_c there will be a belt of no erosion on the given side of the watershed line having a width

$$w_v = x_c \sin A.$$

This marginal belt of no erosion $aa'cc'$ is relatively permanent. It is widest, other things being equal, where the direction of overland flow is most nearly normal to the watershed line; this is usually around the headwaters of an exterior divide. The width of the marginal belt of no erosion decreases for a given x_c as the direction of overland flow becomes more nearly parallel with the direction of the divide, a condition which commonly occurs along lateral segments of the main divide surrounding a drainage basin.

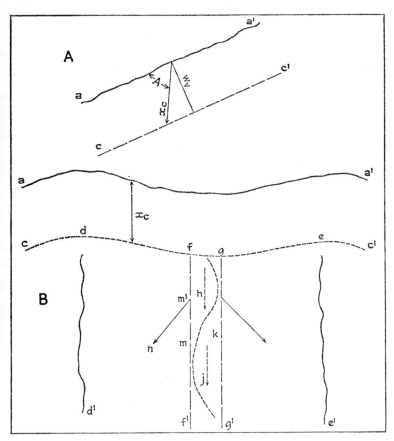

Fig. 5.17 Belts of no erosion.
A, Width of belt of no erosion. B, Longitudinal belt of no erosion.

If aa' (Fig. 5.17B) represents the exterior divide at the head of a newly exposed area, then, with sufficient newly exposed surface, streams will develop, starting at d and e. The entire slope from cc' to the outlet is subject to sheet erosion. Cross-grading begins adjacent to the streams and spreads laterally, until there remains a narrow belt $ff'gg'$ not yet cross-graded. Dashed arrows indicate directions of overland flow antecedent to, and solid arrows the corresponding direction with, cross-grading. This belt has, however, been previously subject to sheet erosion, since it lies downslope from the headward belt of no erosion, and the direction of overland flow is parallel with the slope. The profile of the belt $ff'gg'$ is concave, and it lies, except at its ends, considerably below the original slope. The narrow belt $ff'gg'$ is still subject to cross-grading. Slight variations in surface conditions will divert most of the surface run-off at a given location, as at h, into one stream or the other. The divide between the streams will move away from the stream into which the diversion occurs. The direction of overland flow on the diverted area will swing around until it is more or less parallel with that on the adjacent cross-graded slope, and a belt of no erosion will develop on the side of the divide on which diversion occurs. This belt will have a width $x_c \cos A$, where A is the run-off angle between the diverted surface run-off and the antecedent slope. This angle will vary from zero to $mm'n$, and the width of the belt of no erosion on the given side will vary accordingly.

At some other location, j, the stream ee' will gain the advantage in competition with gg', and the watershed line will be deflected towards ff'. As a result, the watershed line will become sinuous, as shown by the dashed line in Fig. 5.17B. Intermediate between h and j the streams will divide the run-off more or less equally. The watershed line will cross the centre of the belt $ff'gg'$, but at this location most of the run-off will have been diverted at h, and there will be less erosion than at either h or j. As a consequence of the competitive development of divides, the width of the belt of no erosion will vary from point to point, governed locally by the slope, the direction of overland flow and the amount of previously undiverted surface run-off originating within the belt of no erosion. The watershed line will be sinuous in plan and profile, and the watershed ridge will be broken up into a series of irregularly spaced hills, often with flat crestal plateaux, and adjacent hill crests will be at about the same elevations. The hills will be separated by saddles, and will be rounded not only as a result of the manner of their development by aqueous erosion but also by secondary processes, such as earth-slips and rain-impact erosion.

A favourable location for flat-top, interfluve hills is at the junction of a longitudinal and cross-divide. Such junctions commonly occur where there is an angle or bend in the parent divide. Under these conditions the flat-top hill usually has an arm extending out on to

Erosional Development of Streams

the interior divide. Flat-top hills and plateaux may also occur at intermediate locations where there is a relatively wide belt of no erosion.

In Fig. 5.18, aa' and bb' are adjacent tributary streams which developed more or less simultaneously on the same side of the

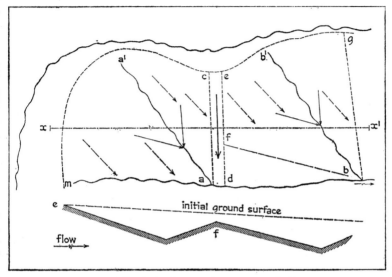

Fig. 5.18 Belt of no erosion at a cross-divide.

parent stream and which flow nearly parallel, and cross-wise of the original slope. When these streams have developed on an antecedent slope, cross-grading will occur, spreading laterally on both sides of each stream. Dashed arrows show directions of overland flow on the antecedent surface, and solid arrows show the corresponding directions after cross-grading by the streams aa' and bb'. Most of the surface run-off on the area $aa'm$ will be diverted from the parent stream into stream aa' by cross-grading.

Downslope from aa' this stream can divert only the run-off from the area $aa'c$. Overland flow on the area upslope from the watershed line ac will be parallel to this line, while on the area $acde$ the antecedent direction of overland flow will still persist, and a belt of no erosion will develop. The stream bb' will receive the run-off from the area $bb'f$ on the upslope side and from the area $bb'g$ on the downslope side. The areas aca' and bgb' will have been subject to at least two cross-gradings, and as a result the direction of overland

F

flow on these areas will have been turned nearly through a right-angle. The direction of overland flow on such areas downslope from streams running crosswise of the original slope may, of course, have either a downslope component (Fig. 5.19), or it may have its direction of flow reversed with respect to the original slope. This happens if the direction of overland flow is deflected through more than 90°.

A longitudinal section along the line xx' is shown in Fig. 5.18. The initial surface is shown by a dashed line. In spite of the fact that e is higher than f, the resultant slope is not materially different on the wide and narrow sides of the valley, a fact often noticed on topographic maps.

It has been shown that the occurrence of a belt of no erosion along an interior divide between streams parallel with the original slope is contingent on the development of components of flow across the divide by micropiracy. A belt of no erosion usually occurs on each side of the divide but is relatively narrow in relation to x_c, and in the vicinity of saddles between crestal hills it may have been subject to cross-grading during its development. If a divide runs crosswise of the drainage basin (*acde*, Fig. 5.18), the belt of no erosion will temporarily be subject to longitudinal erosion, but presently, as a result of erosional competition, hills and saddles will develop, breaking up the longitudinal components of overland flow into elements each less than x_c, as in the case of a longitudinal divide. A crosswise belt of no erosion will usually be wider, and the interfluve hills and plateaux developed thereon will usually be larger and with flatter tops than in case of a divide running parallel with the original slope.

As a drainage system develops, additional belts of no erosion are introduced along the new interior divides, thereby reducing the portion of the total area over which sheet erosion can occur, other things equal. These later divides have been longer subject to gradation than those developed earlier, and they are generally at lower elevations relative to the original surface. Streams that ultimately become the higher-order streams of the drainage basin usually develop early in the erosion cycle, and their divides are usually higher relative to the original surface than those of lower-order tributaries.

A belt of no erosion, once developed, persists throughout subsequent stages of gradation, although subject to variations in width with subsequent cross-grading of the adjacent terrain.

If the drainage basin of a tributary is narrow and steep on one side, with overland flow at right-angles to the divide, the belt of no erosion may extend from the watershed line to the stream on that side, while a flatter slope or overland flow at an acute angle on the opposite side may permit erosion over all or a part of the area on that side.

Discussion thus far has related chiefly to the earlier stages of gradation of a drainage basin, where the length of overland flow is generally much greater than the critical length x_c. At later stages of stream development the critical length x_c is decreased by successive subdivisions, with the birth of new tributaries, until finally there remains little or no intermediate length of overland flow between the belts of no erosion and the streams.

Concordant stream and valley junctions

A new stream develops on a pregraded slope extending away from the parent stream. Thus the new stream enters the parent stream concordantly. A tributary valley has, in general, steeper side-slopes and a smaller value of x_c than its parent stream valley. Hence a tributary valley usually grades faster than the coincident gradation of its parent valley, and although younger, if its stream does not initially enter the parent stream concordantly, it ultimately reaches the grade of the parent stream and debouches into it concordantly. If unrestricted, the tributary stream would cut below the level of its junction with the parent stream, but since it cannot discharge below the grade of the parent stream at the junction and can easily maintain its grade at the level of the latter, it continues to discharge into the parent stream concordantly, as stated by Playfair's law.

Stream-valley gradation

Stream and valley gradation are closely related. The stream supplies a means of disposal of eroded material from the valley and fixes the minimum level of valley gradation. The valley tributary to the stream supplies the run-off that grades the stream. Stream and valley gradation proceed together, but valley gradation tends to lag behind stream gradation.

Fig. 5.19 represents the cross-section typical of a mature tributary stream valley produced by aqueous erosion. As a result of cross-grading and re-cross-grading, the initial surface aa' was cut down to bfh when the stream developed. At each side is a belt of no erosion. For homogeneous material, valley side-slope erosion will not stop at the line bc. If erosion continues until the profile on the left-hand side is bd, the sheet flow, charged with eroded material, arriving at d must be disposed of. Slope is required to carry the

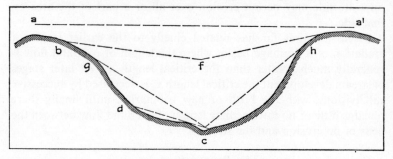

Fig. 5.19 Gradation of stream valley.

water from d to c, and the segment dc will not be graded below this minimum slope. A steeper slope may be maintained from d to c because of sedimentation, if the sheet flow from above d is overcharged with sediment with respect to its transporting power at the reduced slope dc.

For turbulent flow the critical length x_c varies inversely as the surface run-off intensity q_s. Rainstorms range from those with intensities less than infiltration capacity, and which produce no surface run-off, up to the maximum intensity possible in the given locality. In lighter storms x_c will extend to the stream at c, and no erosion will occur, although there may be run-off. In moderate storms x_c will extend to some point between b and c, and erosion will occur only on the lower portion of the slope. Only in maximum storms will x_c be limited to ab, with consequent erosion or sedimentation throughout the length bc.

As a result of the combination of (1) decreasing frequency of higher rain intensities, (2) the existence of a marginal belt of no erosion, and (3) inhibition of valley-bottom erosion by limited transporting power of overland flow, the valley cross-section takes on an ogive or S-shaped form, with a point of contraflexure at some point g and a valley cross-section below this point commonly nearly parabolic in form.

REFERENCES

BAILEY, R. W. (1935) 'Epicycles of erosion in the valleys of the Colorado Plateau Province', *J. Geol.*, XLIII 337–55.
—— FORSLING, C. L., and BECRAFT, R. J. (1934) *Floods and Accelerated Erosion in Northern Utah* (Washington, D.C., U.S. Dept. Agric.) Misc. Pubn. 196.
BEUTNER, E. L., GAEBE, R. R., and HORTON, R. E. (1940) 'Sprinkled plat runoff and infiltration experiments on Arizona desert soils', *Trans. Amer. Geophys. Union*, 550–8.
COTTON, C. A. (1935) *Geomorphology of New Zealand*, I 57.
DAVIS, W. M. (1909) *Geographical essays* (Boston, Ginn & Co.) 777 pp.
DULEY, F. L., and KELLY, L. L. (1941) *Surface Condition of Soil and Time of Application as Related to Intake of Water* (Washington, D.C., U.S. Dept. Agric.) Circular 608.
GILBERT, G. K. (1914) *The Transportation of Debris by Running Water* (Washington, D.C., U.S. Geol. Survey) Prof. Paper 86.
GRAVELIUS, H. (1914) *Flusskunde* (Berlin, Goschen'sche Verlagshandlung) 176 pp.
HORTON, R. E. (1932) 'Drainage basin characteristics', *Trans. Amer. Geophys. Union*, 350–61.
—— (1933) 'The role of infiltration in the hydrologic cycle', ibid., 446–60.
—— (1935) *Surface Runoff Phenomena*, Pt. I: *Analysis of the Hydrograph* (Voorheesville, N.Y., Horton Hydrologic Laboratory) 72 pp.
—— (1937) 'Hydrologic Interrelations of water and soils', *Proc. Soil Sci. Soc. Amer.*, I 401–29.
—— (1938a) 'The interpretation and application of runoff plat experiments with reference to soil erosion problems', ibid., III 340–49.
—— (1938b) 'Rain-wave trains', *Trans. Amer. Geophys. Union*, 368–74.
—— (1939) 'Analysis of runoff plat experiments with varying infiltration capacity', ibid., 693–711.
—— (1940) 'An approach toward a physical interpretation of infiltration capacity', *Proc. Soil Sci. Soc. Amer.*, V 399–417.
—— (1941) 'Sheet erosion, past and present', *Trans. Amer. Geophys. Union*, 299–305.
JEFFREYS, H. (1925) 'The flow of water in an inclined channel of rectangular section', *Philos. Mag.*, 6th ser., XLIX 793–803.
WOOLDRIDGE, S. W., and MORGAN, R. S. (1937) *The Physical Basis of Geography*. (London, Longmans, Green) 435 pp.

APPENDIX: SYMBOLS

A area of drainage basin (sq. miles)
A direction of overland flow
A run-off angle between diverted surface run-off and former slope
α slope angle
c distance from stream tip to watershed line
δ depth of sheet flow, in inches, at the stream margin or at the foot of a slope of length l_o

δ_a	average depth of surface detention of overland flow, in inches, on a unit strip of length l_o
δ_x	depth of sheet flow, in inches, at a distance x from the crest of the slope or watershed line
D_d	drainage density: average length of streams per unit area
l	energy expended by frictional resistance on soil surface, (ft-lb. per sq. ft per sec.)
f	infiltration capacity at a given time t from the beginning of rain (in. per hour)
f_c	minimum infiltration capacity for a given terrain
f_o	initial infiltration capacity at the beginning of rain
F_1	erosive force of overland flow (lb. per sq. ft)
F_s	stream frequency: number of streams per unit area
i	rainfall intensity (usually in. per hour)
I	index of turbulence: percentage of area covered by sheet flow on which flow is turbulent
κ	a proportionality factor which determines the time t_c required for infiltration capacity to be reduced from its initial value f_o to its constant value f_c
κ_l	corresponding coefficient (to κ_s) in the equation for laminar overland flow
κ_s	constant of proportionality factor in equation expressing run-off intensity in terms of depth δ of overland flow
l_g	maximum length of overland flow on a given area
l_o	length of overland flow: length of flow over the ground surface before the run-off becomes concentrated in definite stream channels
$l_1, l_2, ...,$	average lengths of streams of first, second, ..., orders
L'	extended stream length measured along stream from outlet and extended to watershed line
L_o	total length of tributaries of order o
M	exponent in the equation $q_s = K_s \delta^M$, expressing the run-off intensity in terms of depth of sheet flow along the stream margin
n	surface roughness factor, as in the Manning formula
N_o	number of streams of a given order in a drainage basin
N_s	total number of streams in a drainage basin
o	order of a given stream
q_s	surface run-off intensity (usually in. per hour)
q_1	run-off intensity in cu. ft per sec. from a unit strip 1 ft wide and with a slope length

Erosional Development of Streams

ρ	stream-length ratio/bifurcation ratio $=r_l/r_b$
r_b	bifurcation ratio: ratio of average number of branchings of streams of a given order to that of streams of the next lower order. Usually constant for all stream orders in a given basin
r_l	stream length ratio: ratio of average length of streams of a given order to that of streams of the next lower order
r'	stream-length ratio using extended stream lengths
r_s	ratio of channel slope to ground slope for a given stream or in a given basin $=s_c/s_g$
R	hydraulic radius
R_i	initial surface resistance to sheet erosion (lb. per sq. ft)
σ	supply rate $=i-f$
s	order of main stream in a given drainage basin
s_c	channel slope
s_g	resultant slope of ground surface of area tributary to a given parent stream
S	surface slope: fall/horizontal length
t	time from the beginning of rain (hours)
t_l	duration of rainfall excess: time during which the rainfall intensity exceeds infiltration capacity
v	mean velocity of overland flow (ft per sec.)
v	mean velocity of overland flow at the distance x from the watershed line
w_1	weight of run-off, including solids in suspension (lb. per cu. ft)
w_b	width of marginal belt of no erosion $=x_c \sin A$, where A is the angle between the direction of the divide and the direction of overland flow at the divide
x_c	critical length of overland flow: distance from the watershed line, measured in the direction of overland flow, within which sheet erosion does not occur

6 Flood Plains

M. GORDON WOLMAN and
LUNA B. LEOPOLD

A FREQUENTLY quoted definition of a flood plain is that of Rice (1949, p. 139): 'a strip of relatively smooth land bordering a stream [and] overflowed in times of high water'. Valley flats which would usually be considered flood plains, on this definition, include features formed by such processes as landslides and the building of low-angle fans. But the most important process resulting in the formation of valley flats is a combination of deposition on the inside of river curves and deposition from overbank flows. This process has produced many of the flat areas adjacent to river channels, and is probably responsible for most of the flood plains of the great rivers of the world. It is with this particular process that the present paper is concerned.

FREQUENCY OF OVERBANK FLOW

Studies of a number of flood plains in the United States and in India indicate that the frequency of overbank flow is remarkably uniform among rivers flowing in diverse physiographic and climatic regions. Table 6.1 shows the recurrence interval of the bankfull or incipient overflow stage at a number of stations, mainly in the U.S.A. In mountainous areas where broad flood plains are rare, the height of the modern flood plain has been taken as equivalent to the average of the highest surfaces of bars in the channel; this circumstance is in accordance with observations to be presented on the process of flood-plain formation. The frequency with which bankfull height is attained was determined from curves relating discharge or stage of annual floods to recurrence interval (cf. Langbein, 1949). The term *annual flood* is employed in the usual sense of peak discharge in a given water year.

In most of the rivers listed in Table 6.1 the annual flood attains or exceeds the level of the flood-plain surface every year or every other year, giving a recurrence interval of one to two years. Where the flood plain is clearly defined and its elevation accurately known, the

Flood Plains

TABLE 6.1
Recurrence interval with which flood-plain level is attained by annual flood

River and location	Drainage area (sq. miles)	Discharge at bankfull stage (cu. ft.) per second	Recurrence interval, in years (from annual flood series)	Remarks
Pole Creek near Pinedale, Wyo.	88	544	1·13	
Horse Creek near Daniel, Wyo.	124	320	1·10	
Cottonwood Creek near Daniel, Wyo.	202	990	15	Terrace (?)
Little Sandy Creek near Elkhorn, Wyo.	21	227	4	
Big Sandy Creek at Leckie ranch, Wyo.	94	573	1.05	
Green River near Fontenelle, Wyo.	3,970	9,170	1·47	
Hams Fork near Frontier, Wyo.	298	526	1·01	
Middle Piney Creek near Big Piney, Wyo.	34	740	200	Mountain torrent; flood plain poorly defined
Fall Creek near Pinedale, Wyo.	37	1,130	200	Mountain torrent; flood plain poorly defined
Middle Fork Powder River near Kaycee, Wyo.	980	722	1.12	
Red Fork near Barnum, Wyo.	142	80	1·01	
Clear Creek near Buffalo, Wyo.	120	188	1·01	
North Fork Clear Creek near Buffalo, Wyo.	29	390	2·0	Flood plain poorly defined
Little Popo Agie River near Lander, Wyo.	130	534	1·48	
Beaver Creek near Daniel, Wyo.	141	418	1·36	
Rock Creek near Red Lodge, Mont.	100	2,000	11	Mountain torrent; flood plain poorly defined
Yellowstone River at Billings, Mont.	11,870	21,400	1·05	
Seneca Creek at Dawsonville, Md.	101	1,160	1·07	
Bennett Creek at Park Mills, Md.	63	1,510	1·6	
Linganore Creek at Frederick, Md.	82	2,700	2·7	
Big Pipe Creek at Bruceville, Md.	102	3,690	1·25	Only four years of record
Piney Run near Sykesville, Md.	11	?585	1·3	Flood plain difficult to define
Patuxent River near Unity Md.	35	1,330	1·6	
Little Pipe Creek at Avondale, Md.	8	(?)260	(?)1·25	Gauge at distance from flood plain
Brandywine Creek at Chadds Ford, Pa.	287	4,570	1·35	
Buffalo Creek at Gardenville, N.Y.	145	3,000	1·01	
Henry Fork near Henry River, N.C.	80	6,900	3·0	Flood-plain surface not clear
First Broad River near Lawndale, N.C.	198	5,900	1·16	

TABLE 6.1 continued

River and Location	Drainage area (sq. miles)	Discharge at bankfull stage (cu. ft. per second)	Recurrence interval, in years (from annual flood series)	Remarks
South Tyger River near Reidville, S.C.	106	3,400	2·2	
South Tyger River near Woodruff, S.C.	174	15,000	+20	Flood plain difficult to define
Middle Tyger River at Lyman, S.C.	68	1,200	1·18	
Tyger River near Woodruff, S.C.	351	13,500	5	
Willimantic River near South Coventry, Conn.	121	1,050	1·15	
Hop River near Columbia, Conn.	76	730	1·04	Flood-plain elevation not accurately determined
Pomperaug River at Southbury, Conn.	75	1,550	1·05	
Burhi Gandak River at Sikandarpur, Bihar, India.	—	(?)24,000	2·05	'Danger stage' assumed from local information to equal bankfull stage; data from series of gauge height observations
Bagmati River at Dhang railroad bridge, Bihar, India	—	146,000[1]	2·2	'Danger stage' assumed from local information to equal bankfull stage; data from series of gauge height observations

[1] Estimated maximum.

recurrence interval of overbank flow lies closer to one than to two years. There are, of course, variations in data of this kind for natural rivers. Most of the major departures from the average may be ascribed to the difficulty of locating representative flood-plain surfaces produced by the process with which we are here concerned. Narrow valleys on mountain torrents and very turbulent flow on these streams may contribute to this condition. Despite the variations, however, the data from the eastern and western U.S.A. and the examples from India indicate a remarkable uniformity in the recurrence interval of overbank flooding.

Criteria for defining the term *flood* in this context vary among themselves, but all definitions in general imply overbank flow. In this, of course, they differ from the definition of flood in the sense of peak discharge, which does not necessarily imply inundation. Despite the variation in the definition of inundating flood, there is some consistency in the frequency of overbank flows. For instance, the U.S. Army Corps of Engineers (1949a, p. 18) found on the Elkhorn river

Flood Plains

of Nevada evidence for thirty-three floods on the trunk stream, and a greater number on feeder streams, during the sixty-six-year period since 1881. Similarly, records on the Yellowstone (U.S.A.C.E., 1949*b*, p. 22) indicate that at least forty-eight floods have caused serious inundation of farmlands and damage to property along the main stem in the sixty-four-year period from 1882 to 1945. That is, floods have occurred with a frequency of approximately one every 1·3 years.

Flood-damage stage, as used by the U.S. Weather Bureau, refers to the water-surface elevation where overflow begins to cause damage. In an unpublished study of the Bureau's data, W. B. Langbein found that the median recurrence interval of the damage stage at 140 stations was 2·1 years (annual series). The term *flood stage* is in most instances used interchangeably with *flood-damage stage*. Its designation at any particular place is sometimes a reflection of local experience, but usually results from an engineering investigation of flood-control needs, and so becomes accepted as an 'official' designation.

Additional data from the records of the U.S. Army Corps of Engineers on the frequency of flood-damage stage are provided in Table 6.2. The median recurrence interval for this group is about two years, a value somewhat higher than we have found in the majority of our direct field observations.

In order to establish the relation between the elevation of the flood-damage stage and the mean elevation of the flood plain, an effort has been made to compare the two at a number of cross-sections for which data are available. These stations are listed in Table 6.3. The flood-damage stage was obtained from records of the U.S. Army Corps of Engineers (1952) on the Mississippi river and of the U.S. Weather Bureau (1941). The mean height of the flood plain at each cross-section was determined from profiles drawn from topographic maps having 5-ft contour intervals.

Profiles of the Mississippi at Greenville Bridge, Natchez, and at Arkansas City clearly show a difference in elevation between the defined flood stage and the mean height of the flood plain. As can be seen by cross-sections, the difference appears to be due to natural levees adjacent to the river bank which raise the stage at which damage begins. The differences between the two stages are given in Table 6.3. In some cases the damage stage may be equal to or below the elevation of the mean flood plain. In general, however, the damage stage is considerably above the level of the natural flood plain.

The recurrence interval of the flood-damage stage, if known, is

TABLE 6.2

Distribution of flood-damage stage at 71 river gauges in Texas expressed as number of examples in various categories of recurrence interval (data from U.S. Army Corps of Engineers)

Categories of recurrence interval, in years (annual flood series)	1·0	1·0–1·5	1·5–2	2–3	3–5	5–8	8
Number of cases in each category	3	7	16	22	12	4	7

TABLE 6.3

Differences between elevation of flood-damage stage and average elevation of natural flood plain

River and location	Drainage area (sq. miles)	Elevation of gauge datum (ft)	Flood-damage stage (ft)	Flood-damage elevation at cross-section (ft)	Mean flood-plain elevation at cross-section (ft)	Difference in feet (+ indicates damage stage is above flood plain)	Recurrence interval of flood-damage stage [1] (annual flood series)
Mississippi River, at Alton, Ill.	171,500	395·5	21	416·5	425	−8	1·2
Memphis, Tenn.	932,800	183·9	34	217·9	214	+4	1·5
Cairo, Ill.	203,940	270·6	44	314·6	315	0	—
Mouth White River	970,100	108·7	44	152·7	145	+8	—
Arkansas City, Ark.	1,130,700	96·7	44	140·7	132	+9	—
Greenville Bridge, Miss.	1,130,800	74·9	48	122·9	118	+5	—
Lake Providence, La.	1,130,900	69·7	37	106·7	98	+9	—
Vicksburg, Miss.	1,144,500	46·2	43	89·2	84	+5	1·3
St Joseph, La.	1,148,900	33·1	40	73·1	71	+2	—
Natchez, Miss.	1,149,400	17·3	48	65·3	53	+12	—
Arkansas River, at Van Buren, Ark.	150,218	372·4	22	394·4	400	−6	1·5
Sacramento River, near Red Bluff, Calif.	9,300	248·2	23	271·2	265	+6	1·7

[1] List of floods and period of record from U.S. Weather Bureau (1941).

given in the last column in Table 6.3. These values are in general accord with those in the preceding Tables 6.1 and 6.2. It is clear, however, from the comparisons of damage stage and average flood-plain elevation that the stage equal to the elevation of the flood plain will occur more frequently because it is lower. If the frequency of flooding, as determined from the analysis of the damage stage, is adjusted to allow for the difference between the elevation of damage stage and average elevation of the flood plain, the resultant frequency is close to the value obtained from the analysis of those locations where the flood-plain elevation was studied by us in the field. In summary, the annual flood (highest discharge in each year) will equal or exceed the elevation of the flood plain nearly every year.

In view of the obvious differences in the amounts of run-off in different areas, it may seem illogical that there should be a uniform frequency at which overbank flow occurs. Some evidence suggests, however, that where the run-off is high, the channel is larger than where it is small. In other words, the larger channel is adjusted to carry the larger bankfull flows, and thus the frequency with which the flood-plain stage is attained may be the same in regions of diverse run-off.

The uniform frequency of overbank flow suggests that overbank deposits may constitute but a small part of the flood plain. If overbank deposition were a significant feature of flood-plain formation, the observed frequency with which flooding takes place would lead one to expect that flood-plain surfaces would in time be built up to much higher levels than they actually attain, and would in consequence be flooded much less frequently than they are. Again, there would be no reason to expect the uniformity which is actually found in the frequency of inundation in diverse regions.

It is necessary to consider now the nature of the deposits making up the flood plain and some possible explanations of the inferred lack of importance of overbank deposition in flood-plain formation.

FORMATION OF A TYPICAL FLOOD PLAIN

The two fundamental types of deposits which make up a flood plain have often been described in the literature (Fenneman, 1906; Melton, 1936; Mackin, 1937; Happ et al., 1940; Challinor, 1946; Fisk, 1947; Jahns, 1947). The first type is the point bar; the second is the overbank deposit previously referred to. The two types may respectively be described as deposits of lateral and vertical accretion. Despite the distinction between the two in the literature, it is often difficult to distinguish between them in the field. In his study of the Mississippi, Fisk (1947) was able to develop criteria for distinguishing various types of deposits by lithology, texture and morphology. He could separate channel, natural levee and backswamp deposits in the flood plain. On other rivers, however, distinction is more difficult. Modes of deposition, like the deposits themselves, vary. Overbank deposits consist basically of material deposited from high water flowing, or standing, outside the channel. Point bars originate within the channel. Despite the difficulty of distinguishing between these two types of deposits, the distinction is a useful one.

THE CHARACTERISTICS OF POINT BARS

Sediments forming a typical point bar

A point bar is formed on the inside of a river bend by lateral accretion. Deposition is related to the flow associated with the channel bend. Deposition on the convex bank and the concomitant erosion of the concave bank both tend to be greatest just downstream from the position of maximum curvature (see Mackin, 1937, p. 827; Eardley, 1938; Fisk, 1947, p. 32; Dietz, 1952). Together the processes of erosion and deposition tend to maintain a constant channel width during lateral shifts of the channel.

Fig. 6.1 is a detailed map of a point bar on a meander of Watts Branch near Rockville, Maryland. At this point the stream drains about 4 sq. miles and its flood plain is about 270 ft wide. This flood plain, typical of many rivers in the eastern United States, illustrates the type of deposition and stratigraphy commonly found in this area.

The meandering channel contains the flow at low and moderate discharges. At flows large enough to recur only two or three times a year, water passes directly from A to B across the point bar. In this meander we have measured velocities of 3 ft per sec. in water flowing over the point bar, and this velocity is sufficient to scour and to move coarse sand across the bar. Some swale-and-ridge topography on the surface of the bar is produced by these flows. During periods of high flow when rapidly moving water crosses the channel from the convex point towards a position on the concave bank downstream from the position of maximum curvature (from C to D in Fig. 6.1), an eddy usually forms close to the convex bank (E), and is a locus of deposition of relatively fine material. We have often noticed another eddy on the outer, or concave, bank just upstream from the point of maximum curvature of the bend (F). The alternation of high and low flow and the concomitant shifts in the velocity and streamline pattern at any given place give rise to considerable heterogeneity in the point-bar deposits.

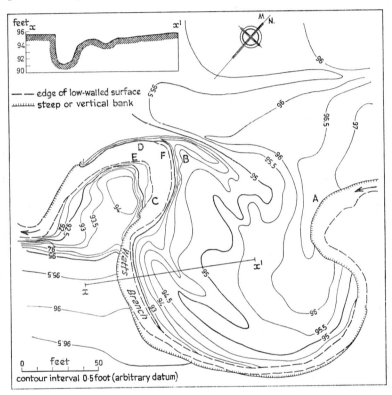

Fig. 6.1 Map and section of a typical point bar on the flood plain of Watts Branch, near Rockville, Maryland.

The map of Watts Branch and the data in Fig. 6.2 demonstrate this diversity. The stream is gradually altering its position in the flood plain (see Fig. 6.4). The nature of the deposits in the point bar is illustrated by Fig. 6.3 (for location, see Fig. 6.2). Fine-grained material was deposited at high flow near the downstream edge of the point bar (sample 1) as a result of small eddies to the left of the main current. The samples at points 2, 4 and 6 include material deposited in slack water adjacent to the main thread of the current. The bimodal character of these deposits appears to indicate that each sample contained material from more than a single sedimentation unit. Most of the stream bed is covered with coarse gravel, and comparable

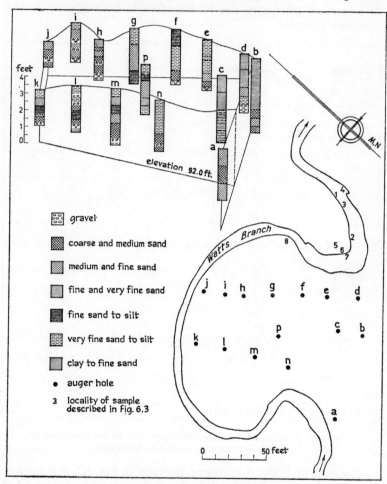

Fig. 6.2 Distribution of materials making up point bar in Fig. 6.1.

material appears in some of the samples in Fig. 6.3. As can be seen in Fig. 6.2, gravel occurs interbedded with sand and with mixtures of clay to fine sand. It is believed that these fine materials were deposited by low or moderate flows. Unless a special effort is made to sample each period of flow separately, a sample will include deposits from two periods. Just downstream from sample 1, the sloping margin of the point bar was covered with piles of leaves and other organic material and generally surrounded by fine mud.

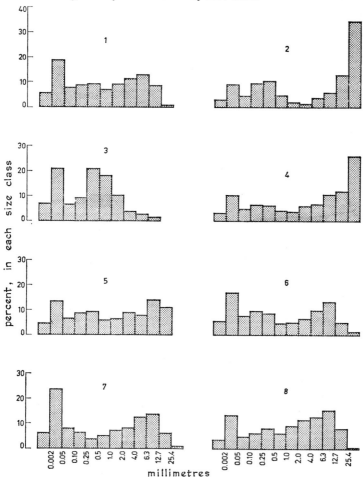

Fig. 6.3 Sample-size distribution from bed and near-bank of flood-plain material of Watts Branch, near Rockville, Maryland.

These examples from active depositional environments illustrate the wide variety of materials constituting the point bar and indicate that, contrary to the common supposition, the point bar is not necessarily composed of material coarser than that which is found in the overbank deposits.

Formation and surface elevation of point bars

During three years of observation and measurement of this point bar on Watts Branch, we have recorded as much as 6 ft of lateral movement of the channel. Cross-sections presented in Fig. 6.4 show that

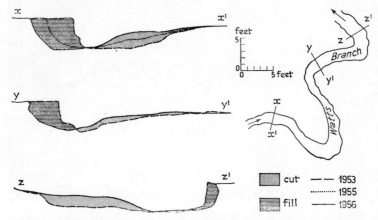

Fig. 6.4 Progressive erosion and deposition on sections of Watts Branch.

erosion of the steep or outside bank was accompanied by deposition on the opposite point bar. At cross-section xx' the amount of material deposited is roughly equivalent to the amount eroded. At yy' erosion has exceeded deposition, while further downstream at zz' during the period of observation deposition has exceeded erosion. We would not expect precise equivalence of erosion and deposition at an individual cross-section during a short period of time. It is clear, however, that the eroding flood plain is being replaced by lateral or within-channel deposits.

The map of Watts Branch presented in Fig. 6.1 might alone suggest that the surface of the point bar is generally lower than the flood plain. Our field observations on many rivers lead us to conclude that the point bar is actually built up to the level of the flood plain; that is, the so-called lateral deposit can be found at all elevations from the water surface to the flood-plain surface. The distinction between overbank and lateral deposits becomes quite vague at stages of flow during which water covers the entire point bar. Overbank flow, as used in this study, refers to stages above the level of the cut bank of the existing flood plain. At times deposition probably takes place in the relatively deep water over the sloping nose of the point bar and in eddies adjacent to threads of rapidly flowing water. Such deposition thus occurs well within the channel. The water flowing over the higher parts or surface of the point bar may scour or deposit, and we have observed both at various times during flood flow over the point bar pictured in Fig. 6.1. Examples from several rivers are presented below which illustrate the accordance of elevation of the point bar and eroding flood plain.

At Watts Branch, stations p, m and n, on the larger of the two point bars in Fig. 6.2, are at an elevation very close to that of the flood-plain surface. On Seneca Creek near Prathertown, Maryland, different point bars composed of a wide variety of material are found to have surface elevations ranging successively from several feet below the flood-plain surface to the height of the surface itself.

The accordance of elevation of a point bar and an adjacent broad flat flood plain was observed on the Burhi Gandak river about 2 miles downstream from Patna in India. A bend in the river had moved laterally about 600 ft in two years. This rapid movement had been carefully noted because the artificial levee being constructed in 1955 was close to the bend. The point bar built on the convex shore of this bend was level with the adjacent flood plain, and near the channel its profile dropped off in a smooth curve, convex upwards, to the edge of the low water surface.

An exceedingly flat-topped gravel bar deposited in the New Fork river near Pinedale, Wyoming, has a composition identical to that found in the flood plain which borders the stream in this reach. Furthermore, its surface is at the same level as the adjacent flood plain. Laboratory observations (Leopold and Wolman, 1957) suggest that such a bar develops by deposition within the channel, often as a linear deposit roughly along the centre-line of the channel. As the upper surface of the submerged bar builds up near the water surface, the flat top is developed under planation by currents and eddies when the depth of flow over the bar becomes relatively small. This same accordance of level and similarity of composition of bar and flood plain are also found on the Popo Agie river near Hudson, Wyoming. The flood plains of these streams appear to be composed of coalescent bars each deposited originally within the stream channel. Although overflow across the bars tends to leave a veneer of fine material on the gravel surface, this veneer makes up a very small part of the total flood plain. In our study of point bars on Little Pipe Creek near Avondale, Maryland, the intimate relation between what is clearly a point bar and the level flood-plain surface was shown by a gravel lens continuous from the point bar into the adjacent flood plain.

In summary, as a stream shifts laterally, deposition on the convex bank keeps pace with erosion of the opposite concave bank. Our data indicate that the surface of the material deposited approaches the elevation of the older part of the flood plain, and thus the volume of material in the point bar is about equal to the quantity of material eroded. Ultimately, all of the debris passing the mouth of a river consists of material eroded from the land of the drainage basin above, but only part of this eroded material moves continuously and directly from its source to the river mouth. Another part is stored temporarily in point bars and in the flood plain at various places in the channel system. Bank erosion of flood plains consists of removal of this material from temporary storage. Point-bar building consists of placing a similar quantity of material into storage.

Our observations indicate that as much as 80 to 90 per cent of a normal flood plain may be composed of deposits of lateral accretion, the remaining 10 to 20 per cent consisting of overbank deposit.

Rate of lateral migration

It is extremely difficult to get reliable data on the normal rate of lateral migration in rivers. Table 6.4 gives a list of measured and estimated rates of lateral migration from a number of sources. These are at best crude. They demonstrate, more than anything else, the variability of lateral movement. Although the larger streams tend to have the more rapid rate of migration, the data show no consistent rates of lateral movement related to the size or position of the channel. Moreover a stream may maintain a reasonably stable position and have but little lateral movement over a long period of time, and then experience very rapid movement during a succeeding period. Considerable error might result from extrapolation of rapid small-scale shifts to long-term movements in a constant direction across an entire flood-plain surface. Even the slower rates shown in Table 6.4, when considered over periods of 500 to 1,000 years, would permit the rivers to move readily from one side of the valley to the other.

OVERBANK DEPOSITION

Nature and amount

There is adequate evidence to show that, in some places and at some times, significant amounts of material are deposited on flood plains by overbank flow. We shall now consider some specific examples.

Flood Plains

At Prentiss Landing on the Mississippi opposite the mouth of the Arkansas river, the Mississippi exposed in 1955 a section containing an old courthouse buried by the river in 1865. The flood plain at this site contains a basal section of cross-bedded coarse sand and silt, constituting channel or point-bar deposits 50 ft thick. These are capped by about 4 ft of finer, banded, overbank deposits; part of these are probably natural levee material, so that the overbank section may be somewhat thicker than normal. Even so, the proportion of the total section made up of overbank deposits is small.

Table 6.5 presents some data on the average thickness of sediment deposited on flood plains by great floods.

Because of the magnitude of these record floods, it is impossible to specify exactly their recurrence interval. From the known record, the Ohio flood of 1937 certainly exceeded a hundred-year flood. It must be recognised, of course, that the figures in Table 6.5 are averages. Under special conditions a foot or several feet of material were deposited. In other places scour rather than deposition occurred. From the data presented by Mansfield (1939, p. 700) the record flood of 1937 on the Ohio river deposited material which would amount to about $\frac{1}{8}$ in. if spread uniformly over the area flooded. However, certain areas were subject to scour, not deposition, and the amount removed was approximately one-quarter as much as the total of material deposited.

In his study of the Connecticut valley, Jahns (1947) describes deposits of the record floods of 1936 and 1938 (Table 6.5) and Pleistocene terrace sequences composed of flood-plain materials. For the most part the overbank deposits of the modern floods ranged in thickness from 6 ft near the stream to a thin veneer at the margins. Jahns (p. 85) estimated that a blanket of sediment was deposited over the entire flooded area to an average depth of $1\frac{3}{8}$ in. during March 1936, and that about $\frac{7}{8}$ in. was added during the hurricane flood of September 1938. The terraces described by Jahns consist of channel and overbank deposits, although he uses different terms for these features. He (p. 49) identified overbank deposits primarily by their stratification and finer texture. The thicknesses of channel and overbank deposits appear in the ratio of about 2 to 1; this is a somewhat larger proportion of material derived from overbank flows than we have found elsewhere. The difficulty of distinguishing point-bar from overbank deposits may account, in part at least, for the difference.

Happ et al. (1940) and many others have noted large amounts of

TABLE 6.4
Some data on rates of lateral migration of rivers across valleys

River and location	Approximate size of drainage area (sq. miles)	Amount of movement (ft)	Period of measurement	Rate of movement (ft per year)	Remarks bearing on measurement and amount of movement	Source of information
Tidal creeks in Massachusetts	±1	0	60-75 years	0		Goldthwait, 1937
Normal Brook near Terre Haute, Ind.	4	30	1897-1910	2-3	Average movement down-valley	Dryer and Davis, 1910
Watts Branch near Rockville, Md.	4	0-10	1915-55	0-0.25		Topographic map and ground survey
Watts Branch near Rockville, Md.	4	6	1953-6	2	Maximum movement; locally in down-valley direction	Successive plane-table surveys
Rock Creek near Washington, D.C.	7-60	0-20	1915-55	0-0.50		Topographic map and ground survey
Middle River near Bethlehem Church, Staunton, Va.	18	25	10-15 years	2.5		Local observer
Tributary to Minnesota River near New Ulm, Minn.	10-15	250	1910-38	9	Tributary near railroad	U.S. Army Corps of Engineers map and aerial photographs
North River, Parnassus quadrangle, Va.	50	410	1834-84	8		Account by local observers
Seneca Creek at Dawsonville, Md.	101	0-10	50-100 years	0-0.20	Maximum age of trees on flood plain is 100 years	Hieb, 1954
Laramie River near Fort Laramie, Wyo.	4,600	100	1851-1954	1	Average movement in ½-mile reach; old map	U.S. Army Corps of Engineers map and aerial photographs
Minnesota River near New Ulm, Minn.	10,000	0	1910-38	0	Most of 10-mile reach	
Ramganga River near Shahabad, India	100,000	2,900	1795-1806	264.	Movement to west; drainage area only approximate	Central Board of Irrigation, 1947
Ramganga River near Shahabad, India	100,000	1,050	1806-83	14	Movement to east; same bend as above	Central Board of Irrigation, 1947
Ramganga River near Shahabad, India	100,000	790	1883-1945	13	Movement to west; same bend as above	Central Board of Irrigation, 1947
Colorado River near Needles, Calif.	170,600	20,000	1858-83	800	One bend (maximum movement in short period)	Means[1]
Colorado River near Needles, Calif.	170,600	3,000	1883-1903	150	One bend (maximum move-	Means[1]

Flood Plains

Yukon River at Kayukuk River, Alaska	320,000	5,500	170 years	32	From evidence furnished by forest succession	Eardley, 1938
Yukon River at Holy Cross, Alaska	320,000	2,400	1896–1916	120	Local observer	Eardley, 1938
Kosi River, North Bihar, India		369,000	150 years	2,460		Ghosh, 1942
Missouri River near Peru, Nebr.	350,000	5,000	1883–1903	250	Rate varied from 50–500 ft annually	Duncanson, 1909
Mississippi River near Rosedale, Miss.	1,100,000	2,380	1930–45	158	About average movement; movement variable because of variations in bank material	Fisk, 1951, Fig. 9
Mississippi River near Rosedale, Miss.	1,100,000	9,500	1881–1913	630	Maximum movement; cut-off channel not included	Fisk, 1951, Fig. 9

[1] Means, T. H. (1953) 'The Colorado River in Mohave Valley: meanderings of the stream in historic times', U.S. Bureau of Reclamation, unpublished report.

TABLE 6.5

Examples of amounts of deposition on flood plains during major floods

River basin	Date of flood	Average thickness of deposition (ft)	Source of data
Ohio River	Jan–Feb 1937	0·008	Mansfield, 1939
Connecticut River	Mar 1936	0·114	Jahns, 1947
Connecticut River	Sep 1938	0·073	Jahns, 1947
Kansas River	July 1951	0·098	Carlson and Runnels, 1952

overbank deposition from individual floods. Occurrences of thick local deposits, even as coarse as gravel, are not uncommon. For example, Harrod and Suter (1881, p. 136) reported that in the Missouri river flood of 1881 the 'immediate banks were raised for long distances from 4 to 6 ft' between Sioux City and Glasgow, Iowa. Although they also noted deposits from 6 to 12 ft in width within the channel in areas which were 'sheltered from the fierce current', even this report makes no mention of widespread deposits sufficient to raise the general level of the flood plain.

Obviously it is difficult to estimate accurately the thickness of sediment deposited over large areas. Although there are examples of thick local deposits from overbank flows, there are also contrasting examples of local scour (see Davis and Carlson, 1952, p. 232; Breeding and Montgomery, 1954, p. 6). Observations in Connecticut following the disastrous record-breaking flood of August 1955 indicate that deposition was extremely irregular and cannot be considered as if it were uniform (Wolman and Eiler, 1957). In a study of the 1951 flood on the Kansas river, it was found that, in general, damage resulting from *deposition* on agricultural land during floods is extremely low (Wolman *et al.*, 1953). The local nature of thick deposits, the large variation in thickness even within a small area, and the occurrence of scour all suggest that widespread deposition of sediment by major floods is not so well established as a glance at Table 6.5 might imply.

If successive increments of overbank deposition were responsible for building a flood plain, we might perhaps expect to find minute laminations representing the deposits involved; but the data in Table 6.5 indicate that lamination would be so fine that it could hardly be detected, much less sampled. Although we have observed short stringers and very thin layers of sand, or even pebbles, in exposures of the flood plain, they have little or no lateral extent. They are certainly lenses rather than extensive overbank layers.

Although material is distributed in the vertical of a stream of flowing water according to size or settling velocity, examples from Brandywine Creek (Wolman, 1955, p. 18), Watts Branch, Seneca Creek and elsewhere indicate that a vertical gradation in size is virtually impossible to find in an individual flood-plain section. Available data show primarily that flood plains may include materials of quite different sizes. Fig. 6.5 gives examples of flood deposits and flood-plain sediments from several regions, showing that the greatest diversity in size occurs in the basin for which most data are available, the Connecticut river basin. Here the major differences in size are due principally to differences in source material within the basin.

Natural levees
The report of Harrod and Suter (1881) on deposition along the 'immediate banks' leads to a consideration of the importance of

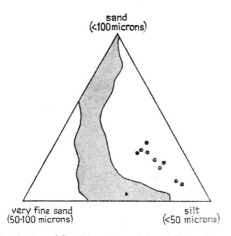

Fig. 6.5 Size distribution of flood deposits and flood-plain sediments: stipple, Connecticut river flood deposits of September 1938 (after Jahns); points, mainly composite analyses for six additional rivers.

natural levees in the formation of flood plains. Many observers have pointed out that when the flow leaves the stream channel its velocity is checked, and as a result the stream is unable to carry its load and deposits material adjacent to the bank. Natural levees are prominent along the Mississippi and Sacramento rivers, where they are particularly well formed along the concave bank at channel bends. Natural levees occur along the Nile river, and are conspicuous on the cross-profiles of many rivers on the Gangetic Plain. Hilgard O'Reilly Sternberg (personal communication) noted that many large alluvial islands in the Amazon river are rimmed by natural levees.

Along the rivers of small and moderate size which exist in Maryland, Virginia and Pennsylvania, we have found few recognisable levees. (See Wolman, 1955, Fig. 16, p. 16, for examples of cross-sections of typical flood plains.) The minor ridges occasionally paralleling the stream not infrequently result from cultivation: farmers ploughing close to a stream often leave a margin of vegetation unploughed, which may trap some sediment.

Natural levees have been described by Happ et al. (1940) in the south-eastern United States. We have also observed very extensive natural levees in Georgia. Because of the coarseness of the material in them, and because of their restriction to certain areas, it is very possible that accelerated erosion related to man's activities has much to do with their formation. Such a possibility is supported by the existence of prominent natural levees on Little Falls Branch and on parts of Rock Creek near Washington, D.C. The levees on these small streams (drainage area 4–50 sq. miles) stand in marked contrast to the absence of levees on most of the streams in the region. Both of the two small creeks drain suburban areas where municipal construction has altered the basin characteristics.

Little Falls Branch not only has natural levees but differs in another way from nearby basins less affected by man's activities. It appears to have experienced a change in frequency of overbank flow. W. W. Rubey (personal communication), who has lived for more than fifteen years in sight of this creek, states that the frequency of overbank flooding has increased from about once a year to at least ten times a year during the period of his observation. The increase can probably be attributed to street construction, paving and home construction on the catchment. It is logical to suppose that open cuts and unpaved streets have during construction greatly increased the amount of sediment carried by the stream. The combination of increased sediment supply and frequency of overbank flow may account for the prominent natural levees.

These illustrations are in accord with Malott's (1928, p. 27) observations that the increase in height of the flood plain adjacent to channels is usually small and noticeable chiefly during low flood stages when this area is the last to be submerged. Although the natural levee is a feature which has received considerable mention in the literature as a type of overbank deposit, it appears to constitute a relatively small proportion of the normal flood plain.

Hypothetical construction of a flood plain by overbank deposition
If a specific thickness of material were deposited on the flood plain every time a river overflowed its banks, it would be possible to compute the rate of rise of the flood-plain surface by the use of flood frequency data. Fig. 6.6 is a plot of hypothetical flood-plain elevation against time on Brandywine Creek at Chadds Ford, Pennsylvania. It was constructed in the following way: The average number of days per year on which a given stage is equalled or exceeded was computed

from the records. Assuming that each time the stream overflows a given level it deposits a specific thickness of material, the time required for the surface of a flood plain to reach a given elevation can be computed. In this example it was assumed that each increment consisted of a layer of sediment 0·005 ft thick (cf. Table 6.5).

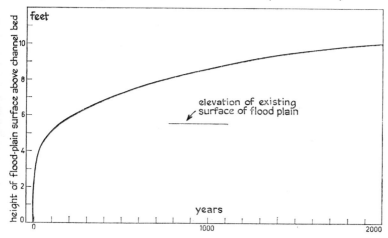

Fig. 6.6 Hypothetical formation of flood plain and increase in flood-plain elevation by overbank deposition: data from Brandywine Creek at Chadds Ford, Pennsylvania.

The most conspicuous feature of the diagram is the rapid increase in elevation in the first ten years. Although the rate decreases thereafter, it would still suffice to construct a flood plain to its present height at Chadds Ford, in a span of 170 years. From 80 to 90 per cent of the construction would be compressed into the first fifty years.

Not far from Chadds Ford is Buck Run, a tributary of Brandywine Creek, near Coatesville, Pennsylvania. In the bank of this tributary channel a log was found at the base of a flood plain. The log lay just below the low-water level, 0·5 ft above the bed of the present stream (see Wolman, 1955, p. 18, Fig. 20). Bedrock lies approximately 3 to 5 ft below the log and crops out in the channel about 50 ft downstream. A carbon-14 analysis dated the log as approximately 1,450 ± 200 years old (Rubin and Suess, 1955, p. 487). The vertical section above the log comprises only 3·1 ft of deposition. The absence of any discernible stratigraphic break in the section allows the possibility that the base of the section is as old as the log. If the log has been in place during this period, then the elevation of the surface of the flood plain has risen so little that it must be considered, in effect, stable.

Assuming that the flood plain at Chadds Ford is about 1,450 years old, we should expect from Fig. 6.6 that the present flood plain would actually be 4 ft above its present position. The fact that it is not suggests that overbank deposition is not the primary mechanism by which a flood plain is formed. Further, if overbank deposition is not a major factor in raising the surface of the flood plain, it follows that in a normal channel the bed will not become progressively farther below its own flood plain as a result of continued overbank deposition.

The data presented thus far indicate that, despite frequent flooding, the elevation of the surface of a flood plain remains stable relative to the level of the channel bed.

Conditions affecting amounts of overbank deposition

Three lines of evidence may help to explain the relative unimportance of overbank deposition in flood-plain formation:

1. Periodic removal of the flood plain by lateral erosion helps to control its height.
2. The highest discharges are often characterised by low concentrations of suspended sediment.
3. Velocities of water on the flood plain during overbank flow may be competent to move sediment of small and medium size.

The extent to which periodic removal and replacement is effective in limiting the height of the flood-plain surface depends on the relative rates of lateral swinging and overbank accretion. The available data do not provide satisfactory comparison of these rates. Some rough relations, however, can be inferred from data presented earlier in this paper.

Table 6.4 indicates that it is likely that specific areas of any flood plain may not be eroded by the stream for periods as long as 200 or more years. Those areas of the flood plain which have not been reached by the river channel for a long time should presumably have higher elevations as a result of continued overbank deposition than the more recently constructed areas of the flood plain. Fig. 6.6 suggests that in 200 to 400 years one could expect 1 to 2 ft of deposition above present flood-plain level as a result of overbank flows. The relief on any flood plain would then be a function of rate of migration or of difference in age. When a flood plain is flat, as many are, the low relief suggests that although the elevation of the flood plain is partially controlled by lateral migration, additional factors also control the amount of overbank deposition.

Contrary to some expectations, high discharges are often associated with lower concentrations of suspended load than are more moderate floods. The curves of sediment concentration and discharge presented in Fig. 6.7 show, for example, that concentration in several Kansas streams first reaches a maximum, and then begins to decline, while discharge is still increasing. In addition, during individual floods the peak of the sediment concentration often precedes or follows the peak discharge.

It has been suggested that the decrease in concentration may result from deposition on the flood plain as a stream goes out of its banks. In the South Fork Solomon river at Alton, Kansas, and in the Solomon river at Beloit, Kansas (Fig. 6.7), the maximum sediment concentration occurs in discharges well below bankfull stage, and thus the decrease of concentration at high flow cannot be attributed to deposition on the flood plain. The fact that high flows may not be associated with the highest sediment concentration may help to account for paucity of deposition during flooding.

Another possible mechanism contributing to small amounts of overbank deposition relates to the ability of overbank flows to transport material across the flood plain. There are many streams transporting material having a moderate range of sizes that do not have natural levees or display a major amount of overbank deposition. Only the finer particles are carried near the water surface or in the upper part of the flow; these sizes are not likely to drop out along the edge of the channel, nor need they be deposited to any great extent on the surface of the flood plain. Water which leaves the channel and flows over the flood plain tends to move directly down-valley, along a slope which may be considerably greater than that of the channel; the higher gradient tends to keep the velocity high and reduces the probability of deposition.

We have assembled records of velocity for overbank sections during inundation, finding them to range as high as 5·4 ft per sec.; and, of the fifty-six velocities determined, nearly half are 2 ft per sec. or greater. These are all mean velocities at particular vertical sections of the overflow. Mean velocities for the whole overflow run lower; of the seven overflows for which such means are determined, however, six give means of 1·2 ft per sec. or greater, and two of more than 2 ft per sec.

These results appear reasonable, if allowance is made for the increase in slope, and also for the fact that roughness of the flood-plain section may be less than the roughness of the main channel.

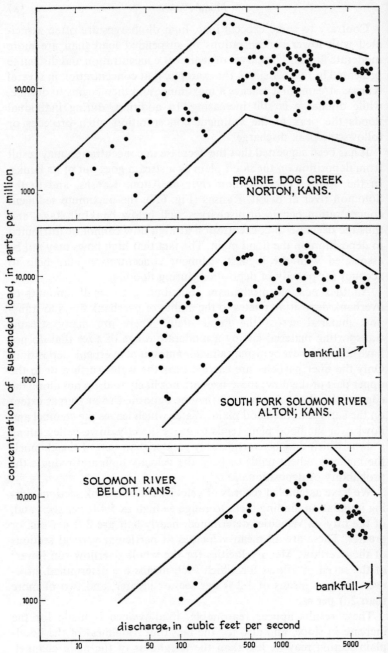

Fig. 6.7 Relation of sediment concentration to discharge for a group of Kansas rivers.

The observed velocities exceed those competent to transport silts and sands (Inman, 1949). Because silt and sand are the predominant sizes carried in the overbank section where such velocities occur, it is perhaps to be expected that much of the material can be carried down-valley and not deposited on the flood plain.

It has been pointed out to us that conditions on the present flood plain may not be representative of the period of flood-plain formation. In particular, the removal of trees from flood plains in the East and in the Middle West may have markedly reduced the likelihood of deposition. In some areas, however, including many in the flood-frequency study previously mentioned, there has probably been little or no change in the plant cover and there is no apparent difference between these and the other flood plains studied.

In summary, we suggest that one possible reason for the general lack of deposition by overbank flow may be the ability of the overbank section itself to transport sediment.

FLOOD PLAINS ON STABLE, AGGRADING AND DEGRADING STREAMS

Thus far the typical flood plain considered here is one in which the relative position of the bed of the channel to the surface of the flood plain has remained stable during the formation of the flood plain. The relatively constant frequency of overbank flooding described earlier could apply even if the entire valley were being aggraded at a constant rate, with channel bed and flood plain rising uniformly. In such an instance there would be the usual difficulty in distinguishing overbank from point-bar deposits, plus the difficulty of distinguishing deposits of different ages in the aggrading sections. The difficulty is illustrated in the following analysis of several flood plains in North and South Carolina and in Georgia. The description of these flood plains is based on numerous borings (Fig. 6.8).

The flood plains rest upon bedrock and have surfaces which are overtopped by the annual flood approximately once each year or once every two years (Table 6.4). They are composed of silt, sand and clay in an infinite variety of combinations. These are commonly underlain by pebble and cobble gravels, as Fig. 6.8 illustrates. From the stratigraphy alone it is impossible to tell how much of this material is overbank and how much is point bar. A log found within a similar flood-plain section in North Carolina is $2,370 \pm 200$ years old.

At first glance it might seem as if these sections are actually aggraded ones rather than stable, alluvial sequences. The following analysis, however, indicates, first, that the distinction between the two is not easily made, and second, that the flood plains in Fig. 6.8 may well represent a single stable deposit.

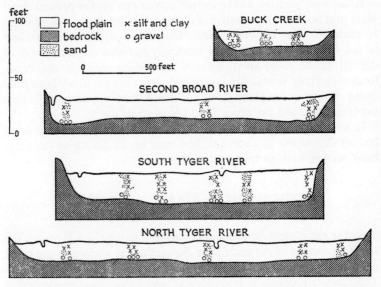

Fig. 6.8 Cross-sections of flood plains in the Carolinas.

Fig. 6.9A is a graph of the depth of fill in valleys in the North and South Carolina region, plotted against the length of the valley from its headwaters to the given cross-section. Although there is considerable scatter on this graph, the average relation can be described by the equation

$$\text{mean depth of fill} \propto \text{length}^{0.45} \quad (6.1)$$

That is, depth of fill is proportional to the 0·45 power of the valley length. Fig. 6.9B relates the mean water depth corresponding to mean annual discharge, in channels of different sizes, to valley length. In many rivers the mean annual discharge is equalled or exceeded about 25 per cent of total time. This second curve, derived from records in the same region that contains the locations in Fig. 6.9A, indicates that mean depth at average discharge is, like depth of fill, proportional to valley length. The relationship is expressed by

$$\text{mean depth of flow} \propto \text{length}^{0.64} \quad (6.2)$$

The progressive downstream increase, alike of depth of fill and mean depth of flow, suggests that the fill may be related to the development of the present channel.

The interrelationship between the flood plain and the present channel is further suggested by computations of the potential depth

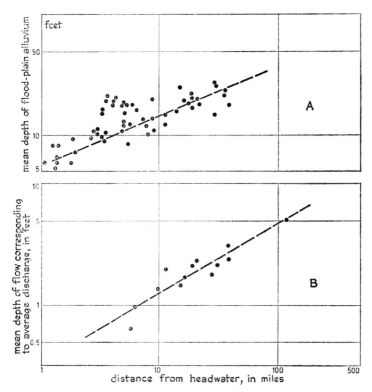

Fig. 6.9 Relation of distance from headwater divide to mean depth of flood-plain alluvium and mean water depth at mean discharge, for localities in the Carolinas.

of scour. Flood records at eight stations indicate that mean depths of flow from 10 to 15 ft (including both scour and depth above mean bed-level) are attained during floods which recur on the average once each five to eight years. Lacey's studies (Inglis, 1949, p. 327) of irrigation canals indicate that the maximum depth of scour in a regime or equilibrium channel may be of the order of 1·75 to 2 times the regime depth. Such scour is associated with a dominant flow in the regime channel which is probably somewhat lower than the flood discharge, with a recurrence interval of about five to eight years. Such flows, however, do not represent extremely unusual conditions. Assuming that conditions in the regime channel are akin to those in natural channels, it is reasonable to suppose that depths of scour $1\frac{1}{2}$ to 2 times the depth attained during these flows are both reasonable and possible. This suggests that the Carolina streams represented in Figs. 6.9A and 6.9B actually scour to bedrock at infrequent intervals. Even if they do so only once every fifty years, this still allows for many scourings during the period of 2,300 years available for the formation of flood plains in the area.

Evidence of scour is also provided by observations of the flood of 1903 (?), which occurred on the Pacolet river in North Carolina. Observers reported that the entire alluvial fill was removed to bedrock by this flood. Allowing for possible exaggerations, it is probably safe to say that scour to bedrock occurred even if the entire alluvial fill was not removed. Because the bedrock is occasionally reached by the flowing water, one must suppose that slow degradation of the valley is taking place, but the rate of degradation is extremely slow: in relation to flood-plain formation, the elevation of the river is in effect stable. Both the frequency of flooding and the calculated potentialities for scour imply that these flood plains are being continuously constructed and reconstructed by the streams which now flow within them.

It has already been pointed out that flood plains can exist in aggrading channels. During aggradation, floods occur and the stream as a rule experiences a certain range in discharge. Both the bed of the channel and the surface of the flood plain continue to rise. Although the channel may be choked with material at times (Happ, 1950), there is no reason to believe that a flood plain will not continue to be constructed during the rise.

Where continual aggradation has produced the valley fill, it is difficult to explain how the relative position of the channel to the flood plain remained fixed during aggradation if overbank deposition is considered the principal mechanism of laying down the valley fill. Rather, concomitant rise of both stream bed and flood-plain surface appears to be best explained by attributing the bulk of the deposited material to the process of point-bar formation.

The uniform frequency of flooding of flood plains does not rule out the possibility that both the surface of the flood plain and the bed of the channel are being built simultaneously. Gauges on the

Nile river, which provide the longest periods of record of any river in the world, indicate that both the bed and the flood plain of the Nile are being raised at a rate of about 3 to 4 ft in 1,000 years (Lyons, 1906, pp. 313–17). Such rates are unmeasurable in the brief periods of record with which we are dealing. The logs dated at 1,500 and 2,300 years old, beneath the two flood plains in Pennsylvania and North Carolina, provide the only available fairly good evidence of the age of modern flood plains. The dates accord generally with datings derived from terrace sequences in the West (Leopold and Miller, 1954).

It must be admitted that the stability of the absolute elevation of the surfaces of most flood plains cannot yet be proved. The evidence demonstrates, however, that even during aggradation the difference in elevation between the river bed and the surface of its flood plain does in many instances remain constant over long periods of time. That the same may be true during periods of rapid degradation as well as the slow degradation previously mentioned is indicated by observations of the Ukak river in Alaska. In 1912 volcanic ash from the eruption of Katmai filled the river valley. In the forty-year period since the eruption the river has cut down 10 to 40 ft and is still continuing to cut; at the same time it has constructed a flood plain and is still engaged in modifying it.

These various observations indicate that a channel may have a flood plain associated with it when it is stable and flowing on bedrock, when it is gradually eroding a valley, or when it is gradually depositing a fill.

When aggraded valley fill or any flood plain is incised after its formation, the former flood plain becomes a terrace. An alluvial terrace is an abandoned flood plain whose surface no longer bears the normal relation to the stream bed. This study indicates that the normal relation between the surface of the active flood plain and the stream bed on many streams can be defined by the frequency of flooding. Where such is the case, a flood plain becomes a terrace when the channel incises itself to the point where the former active flood plain is no longer overtopped by that annual flood which on the average occurs less than once every two years.

Finally, we ought to point out that while point bars are described in the literature primarily in relation to meandering streams, and while many of the examples given here are also taken from such streams, we have shown elsewhere (Leopold and Wolman, 1957) that there are only small differences in the fundamental characteristics of so-called straight channels, meanders and braids. Flood plains may be constructed by channels of any type or size.

CONCLUSION

This study supports the view that the flood plain is composed of channel deposits, or point bars, and some overbank deposits. The relative amounts of each vary, but on the average the proportion of overbank material appears to be small. This conclusion is supported by the uniform frequency of flooding and by the small amount of deposition observed in great floods. Lateral migration, relatively high velocities which can occur in overbank flows, and the decrease in sediment concentration at high flows contribute to this result.

The flood plain can only be transformed into a terrace by some tectonic, climatic or man-induced change which alters the regime of a river, causing it to entrench itself below its established bed and associated flood plain.

REFERENCES

BREEDING, S. D., and MONTGOMERY, J. H. (1954) *Floods of September 1952 in the Colorado and Guadalupe River basins, Central Texas*, U.S. Geol. Survey, Water-Supply Paper 1260-A, 46 pp.

CARLSON, W. A., and RUNNELS, R. T. (1952) 'A study of silt deposited by the July 1951 flood, central Kansas river valley', *Kans. Acad. Sci. Trans.*, LV 209–13.

CENTRAL BOARD OF IRRIGATION, INDIA (1947) *Annual Report (Technical)*, II 552.

CHALLINOR, J. (1946) 'Two contrasted types of alluvial deposits', *Geol. Mag.*, LXXXIII 162–4.

DAVIS, S. N., and CARLSON, W. A. (1952) 'Geological and groundwater resources of the Kansas river valley between Lawrence and Topeka, Kansas, *Kans. Geol. Survey Bull.*, XCVI 201–76.

DIETZ, R. A. (1952) 'The evolution of a gravel bar', *Mo. Bot. Garden Ann.*, XXXIX 249–54.

DRYER, C. R., and DAVIS, M. K. (1910) 'The work done by Normal Brook in 13 years', *Ind. Acad. Sci. Proc.*, 147–52.

DUNCANSON, H. H. (1909) 'Observations on the shifting of the channel of the Missouri river since 1883', *Science*, XXIX 869–71.

EARDLEY, A. J. (1938) 'Yukon channel shifting', *Geol. Soc. Amer. Bull.*, XLIX 343–358.

FENNEMAN, N. M. (1906) 'Floodplains produced without floods', *Bull. Geog. Soc. Amer.* XXXVIII 89–91.

FISK, H. N. (1947) *Fine-grained Alluvial Deposits and their Effects on Mississippi River Activity*, U.S. Waterways Exp. Sta., 2 vols., 82 pp.

—— (1951) 'Mississippi river valley geology: relation to river regime', *Amer. Soc. Civ. Eng. Trans.*, CXVII 667–89.

GHOSH, B. P. C. (1942) *A Comprehensive Treatise on North Bihar Flood Problems* (Patna, Bihar, Govt. Printing) 200 pp.

GOLDTHWAIT, J. W. (1937) 'Unchanging meanders of tidal creeks, Massachusetts' (abstract), *Geol. Soc. Amer. Proc.* (1936) 73–4.

GOLDTHWAIT, R. P. (1941) 'Changes on the intervales of Connecticut and Merrimac rivers', *N. H. Acad. Sci. Proc.*, I 17.

HAPP, S. C. (1950) 'Geological classification of alluvial soils' (abstract), *Geol. Soc. Amer. Bull.*, LXI 1568.

—— RITTENHOUSE, G., and DOBSON, G. C. (1940) *Some Principles of Accelerated Stream and Valley Sedimentation*, U.S. Dept. Agric., Tech. Bull. 695, 133 pp.

HARROD, B. M., and SUTER, C. R. (1881) *Report of the Committee on Outlets and Levees upon the Floods of the Missouri River in the Spring of 1881*, 47th Cong., 1st sess., S. Ex. Doc. no. 10, I 135–9 (App. H).

HIEB, D. L. (1954) *Fort Laramie National Monument, Wyoming*, U.S. Dept. Interior, Natl. Park Service Hist. Handbook, series 20, 43 pp.

INGLIS, C. C. (1949) *The Behaviour and Control of Rivers and Canals* (Poona, Central Water-power Irrigation and Navigation Res. Sta.) Res. Pub. 13, 2 vols., 486 pp.

INMAN, D. L. (1949) 'Sorting of sediments in the light of fluid mechanics', *J. Sed. Petrology*, XIX 51–70.

JAHNS, R. H. (1947) *Geologic Features of the Connecticut Valley, Mass., as Related to Recent Floods*, U.S. Geol. Survey, Water-supply Paper 996, 158 pp.

LANGBEIN, W. B. (1949) 'Annual floods and the partial-duration flood series', *Trans. Amer. Geophys. Union*, XXX 879–81.

LEOPOLD, L. B., and MILLER, J. P. (1954) *A Postglacial Chronology for some Alluvial Valleys in Wyoming*, U. S. Geol. Survey, Water-Supply Paper 1261, 90 pp.

——, and WOLMAN, M. G. (1957) *River Channel Patterns: Braided, Meandering, and Straight*, U.S. Geol. Survey, Prof. Paper 282-B.

LYONS, H. G. (1906) *The Physiography of the River Nile and its Basin* (Cairo, Egypt, Survey Dept.) 411 pp.

MACKIN, J. H. (1937) 'Erosional history of the Big Horn Basin, Wyoming', *Geol. Soc. Amer. Bull.*, XLVIII 813–94.

MALOTT, C. A. (1928) *Valley Form and its Development*, Ind. Univ. Studies, no. 81, 26–31.

MANSFIELD, G. R. (1939) *Flood Deposits of the Ohio River, January–February 1937: A Study of Sedimentation*, U.S. Geol. Survey, Water-supply Paper 838, 693–733.

MELTON, F. A. (1936) 'An empirical classification of flood-plain streams', *Geog. Rev.*, XXVI 593–609.

RICE, C. M. (1949) *Dictionary of Geological Terms* (Ann Arbor, Mich., Edwards Bros.), 461 pp.

RUBIN, M., and SUESS, H. E. (1955) 'U. S. Geological Survey radiocarbon dates. II', *Science*, CXXI 481–8.

U.S. ARMY CORPS OF ENGINEERS (1949a) *Examination Elkhorn River and Tributaries, Nebraska*, 81st Cong., 1st sess., H. Doc. no. 215, 60 pp.

—— (1949b) *Examination Yellowstone River and Tributaries*, 81st Cong., 1st sess., H. Doc. no. 216, 71 pp.

—— (1952) *Stages and Discharges of the Mississippi River and Tributaries in the Vicksburg District*, Vicksburg Dist., 281 pp.

U. S. WEATHER BUREAU (1941) *The River and Flood Forecasting Service of the Weather Bureau* (Washington) p. 16.

WOLMAN, A., HOWSON, L. R., and VEATCH, K. T. (1953) *Report on Flood Protection, Kansas River Basin, Kansas* (Kansas City, Mo., Indus. Devel. Comm.) 105 pp.

WOLMAN, M. G. (1955) *The Natural Channel of Brandywine Creek, Pa*, U.S. Geol. Survey, Prof. Paper 271, 56 pp.

——, and EILER, J. P. (1957) 'Reconnaissance study of erosion and deposition produced by the flood of August 1955 in Connecticut', *Trans. Amer. Geophys. Union* (in press).

7 River Channel Patterns

LUNA B. LEOPOLD and
M. GORDON WOLMAN

FROM the consistency with which rivers of all sizes increase in size downstream, it can be inferred that the physical laws governing the formation of the channel of a great river are the same as those operating in a small one. One step towards understanding the mechanisms of operation is to describe many rivers of various kinds. The present study is concerned principally with channel pattern. The term refers to limited reaches of channel that can be defined as straight, sinuous, meandering or braided. Channel patterns do not, however, fall easily into well-defined categories, for, as will be discussed, there is a gradual merging of one pattern into another. The difference between a sinuous course and a meandering one is a matter of degree. Similarly, there is a gradation between the occurrence of scattered islands and a truly braided pattern.

The interrelationship among channels of different patterns is the subject of this study. Because neither braided channels nor straight channels have received attention in the literature comparable to that given to meanders, we shall begin with them.

THE BRAIDED RIVER

The term *anastomosis* appears to have been first applied to streams by Jackson (1834). It was again used by Peale (1879, p. 528) in a description of channel pattern on tributaries of the Green river in Wyoming. Because the term has occasionally been misapplied in geomorphic writing, it is desirable here to recall its definition – the union of one vessel with another, or the rejoining of different branches which, arising from a common trunk, form a network. Successive division and rejoining with accompanying islands is the important characteristic denoted by the synonymous terms *braided* or *anastomosing* streams.

A braided pattern is probably associated, in many minds, with the concept of aggradation. Not until Rubey (1952) examined the matter was channel division with island formation discussed as one of the possible equilibrium conditions of a channel. The following examples extend Rubey's work.

Horse Creek: type locality of the braided stream
It is appropriate to use as the first example of a typical braided river the stream to which the term *anastomosing* was early applied by Peale. Fig. 7.1 shows the Horse Creek–Green river confluence, as

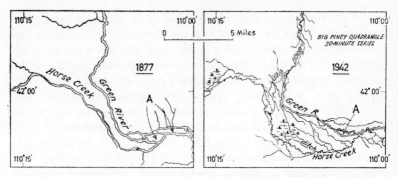

Fig. 7.1 Lower Horse Creek and the Green river near Daniel, Wyoming: left, *from Peale (1879);* right, *from Geological Survey (1942).*

depicted by Peale and as shown on a modern map. The islands drawn in by Peale still exist, changed in form only to a minor extent.

Within a few miles of the point where Peale made his observations, we have mapped a reach of Horse Creek which includes a gauging station (Fig. 7.2). At this place the stream has a drainage area of 124 sq. miles and a mean discharge of about 65 cusecs. The braided pattern, although not so well developed here as farther downstream, is still apparent. This reach was selected for study because the channel pattern is typical, and because the records of discharge can be used to analyse the flow characteristics.

At the gauging station (cable in Fig. 7.2) the stream has a single channel, but the flow divides a few hundred feet downstream, before reuniting once again. The division begins as a low gravel bar (*C* in the diagram) near the left bank, continuing into a central ridge, which in turn meets the tip of a gravel island (*D*). This island supports a willow thicket. The left-hand channel passing the island is itself divided by a linear gravel bar (*E*) which, beginning about 200 ft past the upstream tip of the island, extends for nearly 250 ft along the centre of the channel and ends near the junction at the far end of the island.

River Channel Patterns 199

The left-hand and right-hand channels are about equal in width – 30 to 40 ft across. Their combined width is 20 to 30 per cent greater than the 50-ft width of the undivided reach. The central gravel bar

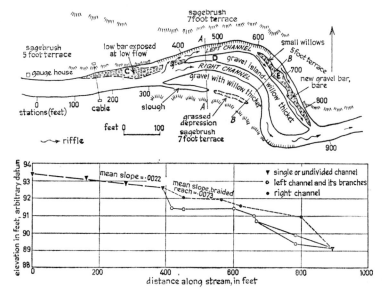

Fig. 7.2 Plan and profile of Horse Creek near Daniel, Wyoming (sections given in Fig. 7.3).

in the left-hand channel is associated with but a slight increase in width there.

Opposite the gravel island is a slough aligned with a grassed depression. Both features undoubtedly carry water during flood flow. Their configuration and their position indicate that they were once parts of a continuous active channel which has been blocked by deposition. In its active state the old channel was separated from the existing right-hand channel by an island, similar to the gravel island between the left-hand and right-hand channels today.

The linear bar in the left channel is gently rounded in cross-section at its upstream end, where it is lobate in plan. In the downstream direction it becomes pointed, flat on top, and trapezoidal in cross-section. It carries little vegetation, except at the farthest upstream end where young willows have established themselves; however, in the year's interval between two series of observations, eighty individual plants ranging through nine species of grasses and wild flowers appeared on the 425-sq.-ft area of open gravel.

G 2

Two cross-sections of the braided reach are given in Fig. 7.3. The valley bottom here is bounded by two low terraces, one 5 ft and the other 7 ft higher than the low-water surface. Between the confining

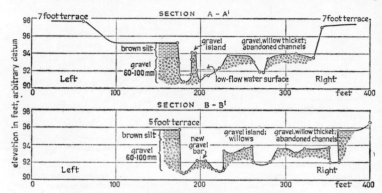

Fig. 7.3 *Cross-sections of braided reach of Horse Creek (for plan and profile, see Fig. 7.2).*

banks of the 5-ft terrace, the gravel island and the adjacent gravel flat containing the abandoned channel lie at a single elevation. The new gravel bar is generally somewhat lower than the adjacent island, although at about halfway along its length the flat upper-surface stand reaches island level.

Stages in the development of a braid

These observations suggest a sequence of events in the development of a braided reach. In an originally single or undivided channel, a short, submerged central bar is deposited during a high flow. The head of the bar is composed of the coarse fraction of the bed load. Because of some local condition not all the coarse particles are transported through this particular reach and some accumulate in the centre of the channel. Part of the finer fraction is trapped by the coarser material and also deposited. Though the depth is gradually reduced, velocity over the growing bar tends to remain undiminished or even to increase, so that some particles moving along the bar near the centre of the channel roll along the length of the new bar and are deposited beyond the lower end, where a marked increase in depth is associated with a decrease in velocity. Thus the bar grows by successive addition at its downstream end, and presumably also by some addition along the margins. Downstream growth is suggested by willows established at the upstream tip while the downstream portions were still bare. A similar gradation in age of vegetation exists in many other islands studied.

When the bar gets large enough, the channels along its sides are insufficient in width to remain stable. Widening then occurs by trimming of the central bar and by cutting of the original sides of the channel until a stable width has been attained. At the same time some deepening of the flanking channels may occur and the bar emerges as an island. The bar gradually becomes stabilised by vegetation. At some stage lateral cutting against the bar becomes just as difficult as against the banks of the original channel, and so the bar is not eliminated. The hydraulic properties of the channel during this process of island formation will be discussed in a later section.

After the island has been formed, the new channels in the divided reach may become subdivided in the same manner. As successive division occurs, the amount of water carried by individual channels tends to diminish, so that in some of them vegetation prevents further erosion and promotes deposition.

The Horse Creek example demonstrates all of these features.

The gravel island separating the two main channels is considerably wider than the new gravel bar. The right channel is separated from the abandoned channel by a former island which was also wider than the new bar. When a central bar is deposited, it may continue to increase in width, forcing the channels farther apart. One reason for this lateral cutting can be seen in Fig. 7.2; the low bar C has built downstream until it actually joins the upstream tip of the gravel island D. At low flow the water which gets into the left channel pours over the downstream tip of the low bar in a direction nearly perpendicular to the general stream course. At high discharge the flow impinges against the left bank and subsequently produces a sharp bend in the streamlines to the right as they become aligned again with the left channel. Thus the low bar and the upper tip of the gravel island force the flow into an S-shaped path. As a consequence the left bank tends to erode where the flow impinges against it, while the inside of this curve is a zone of deposition which blunts or widens the upstream tip of the gravel island. The widening of an initial linear bar is probably due mostly to the deposition on the inside of the bend, resulting from obstruction by the bar itself.

The gravel island is taken for a stabilised and enlarged bar. It has a thin layer of silt which, covering the surface of the underlying gravel, is believed to have been arrested by vegetation during overbank flow. The initial vegetation which sprouts on a new gravel bar begins the screening process, and the consequent deposition of thin patches of silt or fine sand promotes the stand of vegetation. Screening of fine material and the improvement of the stand of vegetation by altering the texture of the surface layer are reciprocal and self-perpetuating.

CHANGES ASSOCIATED WITH CHANNEL DIVISION

Flume experiments

Experimental work in a flume at the California Institute of Technology allowed us to test the hypothesis of bar deposition just outlined. The 60-ft flume had a width of about 3 ft and was filled to a depth of about 5 in. with a poorly sorted medium sand. Initial channels of various shapes and sizes were moulded by means of a template mounted on a moving carriage, the latter carrying a point gauge for measuring elevation to an accuracy of 0·001 ft vertically and 0·005 ft horizontally. The slope of the flume was adjusted by the supporting jacks: slopes of 0·01 could be obtained. The flowing water itself altered the initial channel section and slope. Discharge into the flume system was measured by a Venturi meter. From the meter the water flowed into a stilling basin and thence through a honeycomb for suppressing turbulence. A short trapezoidal channel of wood connected the stilling pool to the alluvial channel. Discharges used were in the range 0·01–0·10 cusecs. At the downstream end of the sand channel was a trapezoidal weir, the elevation of which was kept approximately level with the sand bed of the stream.

Sediment, in the form of dry sand, was fed into the system from a hopper. The rate of feed could be varied from zero to about 200 gm. per minute. During the experiments three sizes of sand were used.

Although the rate of sand feeding could be determined, it was impossible to measure accurately the amount delivered to the settling basin at the lower end. In order to ascertain if the system was in equilibrium, we were forced to rely on an inexact method – successive determinations of the long profiles of water surface and of dry stream bed. This method, however, is sufficiently accurate where the rate of introduction of load is large relative to the channel, as in most of the experimental runs, for aggradation or degradation could be accurately measured during the progress of a run of reasonable duration.

It must be stressed that the channel developed in the flume was not a model river, but the prototype of a small stream. The flume-river adjusted not only its slope, but also its depth and width.

A period during which sand and water were delivered to the channel at constant rates was as long as twenty-five hours, with interruptions at intervals to permit measurement. In the first example showing the development of a braided reach, the initial channel cut by the template was 15 in. wide and 1½ in. deep, with the slope set at 0·0114. The discharge was 0·085 cusecs. Load was introduced at the rate of 120 gm. per minute, giving a sediment concentration of 830 p.p.m. by weight. It frequently happened that the channel width increased very rapidly from its initial value under the action of flowing water, attaining a minimum value which was stable for the particular discharge. In the circumstances outlined, the average stable width was 1·1 ft, so that no rapid adjustment took place. During twenty-two hours of flow, a series of bars and islands developed in a 12-ft reach between stations 10 and 22, beginning 7 ft downstream from the entrance.

The sequence of development in this braided reach is shown in Fig. 7.4. At the end of three hours of flow the development of a central submerged bar had proceeded so far that its lower end had caused some deflection of flow towards the right bank and the appearance of a small arcuate re-entrant in the formerly straight bank. The head of the submerged bar had similarly caused some cutting on the left bank.

The cross-section at station 14 after four hours of flow shows the pronounced central ridge or submerged bar. It should be noted, however, that in the upper left diagram showing the plan view at three hours, the band of principal bed transport lies on top of the submerged central bar. The grain movement in the deeper parts of the channel adjacent to the central bar was usually negligible in the early stage of island development. The central bar continued to build closer to the water surface, yet the principal zone of movement remained for a time along the top of the building bar. It was there that some of the larger particles stopped, trapping smaller particles. These could have been moved had they not become protected or blocked by larger grains.

The central bar in the flume-river was caused by local sorting, the larger particles being deposited in the centre of the stream at some place where local competence was insufficient to move them. But, as explained, bar building by sorting does not imply that the deposit is composed only of grains too large to be moved further. The sketches of the developing braided reach show an irregular distribution of well-sorted fine material, usually deposited near the toe or downstream edge of the bar where progressive bar development consisted of foreset beds as in a delta. The continual shifting of the channels, then, builds a heterogeneous bar consisting of patches of materials of different size and different degrees of sorting.

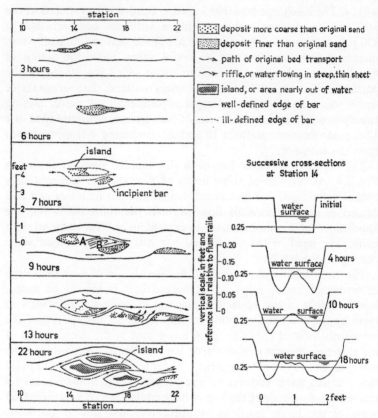

Fig. 7.4 *Development of braids in flume, in plan and in section.*

By the end of seven hours of flow the central bar had been built so high that individual grains rolling along its ridge actually 'broke' the water surface. Leopold and Miller had noted previously (1956, p. 6) that in arroyos, many large pebbles and cobbles were observed to roll in flows of depth only half the diameter of the rolling stone. By such a process a true island cannot be formed in uniform discharge. However, the deflection of the water by the central bar usually caused local scour and a lowering of the water surface which left the bar sticking out of the water as a real island.

The area designated as 'incipient bar' in Fig. 7.4 at hour 7 had enlarged and emerged as an island by hour 9. Furthermore, after hour 9 of flow a linear bar had developed in the channel area between the two islands, bisecting the flow into two parts, marked A and B. By hour 13, continued bar building in channel A and subsequent lowering of the channel B diverted most of the water out of the old channel A, and the primary zone of transport over the bar area was restricted to channel B.

Similar sequences continued to change the configuration of the braided reach, moving the principal zones of transport from one place to another, and at hour 22 there were three island areas. Since in the flume-river the constant discharge did not permit the islands to receive any increment of deposition after emergence, an 'island' actually represents merely the highest knob of a bar most of which remains submerged. The cross-section of station 14 at hour 18 (bottom diagram, right side, Fig. 7.4) illustrates this feature.

The succession of events observed in the flume are analogous to those postulated to account for the characteristics of the reach studied on Horse Creek. The central bar built closer to the water surface and extended itself downstream with time, channels were successively formed and abandoned, and the bars were made up of the coarser fractions of the introduced load but mixed with considerable fine material which had become trapped.

Nearly all observers who have recorded notes on the action of a braided stream have remarked on the shifting of bars and the caving of banks. Once a bar is deposited it does not necessarily remain fixed in form, contour or position. This can be seen in the successive cross-sections of four stations made during the run pictured in Fig. 7.5. Time changes are arranged side by side; downstream variation can be seen in the vertical groups. A reference level is given for each section so that the changes of water-surface elevation can be compared. It can be seen that station 6 underwent continued aggradation and that the shape of the cross-section changed radically – much more than did the area of the cross-section.

The sections for station 10 illustrate the varying elevation of water surface which rose between hours $1\frac{1}{2}$ to 4, fell from hours 4 to 6, and rose again from hours 6 to 11. The fall of water surface elevation is clearly due to the scour of the right-hand channel between hours 4 and 6.

The initial linear nature of the central bar is best illustrated at hour 6, where a central ridge is present through an 11-ft reach between stations 6 and 17. This is shown in the plan and cross-sections in the lower right of Fig. 7.5.

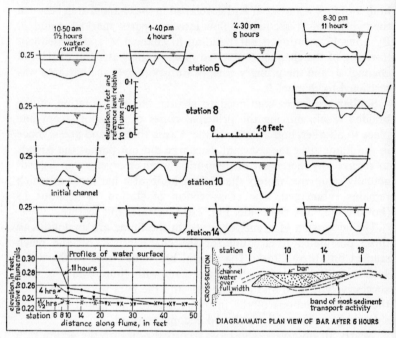

Fig. 7.5 *Cross-sections of bar development in braided flume channel; diagrammatic plan of bar.*

The growth and subsequent erosion of a central bar is also illustrated at station 10. After one and a half hours the initial moulded channel shape had been altered by the building of a central bar and by slight degradation of the channels beside it. By the end of four hours the combination of lateral building and deposition on the bar surface had widened the bar and made a double crest. Bar building resulted in a diversion of most of the water into the right channel at station 10, causing local scour.

Divided and undivided reaches in the flume and in natural rivers
In the natural streams and in the flume-river alike, the slope of a divided reach proves to be greater than that of an undivided reach. The steepening of the divided reach in the flume is very marked

(Fig. 7.6). The initial profile, shown as a heavy line, represents the slope of the water surface when the run began. By the end of nine hours aggradation had taken place in the reach between station 6

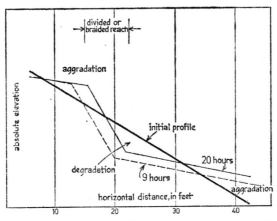

Fig. 7.6 Diagrammatic profiles interpreting measurement data.

and station 12, and also downstream from station 35. Degradation of the initial channel had occurred between stations 15 and 35 with the establishment of a steep reach in the divided or braided section.

Between hours 9 and 20 of flow, continued aggradation took place in the divided reach, but in general the steep slope was maintained approximately parallel to that which existed at the end of hour 9. Similarly the aggrading reach downstream from station 20 maintained nearly the same average slope as that which existed at hour 9, this slope being much flatter than that of the divided or braided reach. A similar sequence can be observed in the profiles in the lower left of Fig. 7.5: between hours $1\frac{1}{2}$ and 4 the reach from station 10 to station 22 steepened markedly, but from hours 4 to 11 aggradation occurred at approximately the same slope which existed in this downstream reach at hour 4. In the braided reach upstream from station 10, however, continued aggradation was accompanied by continued steepening between hours 4 and 11.

A most important observation in these experiments is that aggradation could take place at a constant slope, and without braiding, even when the total load exceeded the capacity of the channel for transport. Braiding is developed by sorting as the stream leaves behind those sizes of the load which it is incompetent to handle. If such sorting results in progressive coarsening of the bed material, then the slope increases progressively. If the stream is competent to move all sizes comprising the load but is unable to move the total quantity provided to it, then aggradation may take place without braiding. Hence, contrary to the assumption often made, the action in the flume-river suggests that braiding is not a consequence of aggradation alone.

Although the steepening of slope in the divided reach is one of the more obvious of the observed changes in the channel associated with division around an island, nearly all other hydraulic parameters are also affected. Detailed comparisons of an undivided channel and a channel divided around an island, or a potential island, are available for three river reaches and four runs in the flume-river. These comparisons are presented in Table 7.1. Comparisons are shown as ratios of the measurements in the divided reach to similar measurements in the undivided one. The width in a reach containing an island is the width of flowing water. The width of the undivided reach is the width of the water surface upstream or downstream from the island of where there is but a single channel.

All examples show that channel division is associated with increased width of water surface, increased slope and decreased depth. In the three comparisons from our measurements of natural rivers, the sum of the widths of the divided channels ranges from 1·6 to 2·0 times that of the undivided one. In the four comparisons made in the flume, the ratios vary from 1·05 to 1·70.

This increase in water-surface width caused by development of a bar or an island is accompanied by a decrease in mean depth. The ratio of depths in the divided reach to depth in the undivided reach varied from 0·6 to 0·9 in natural rivers and from 0·5 to 0·9 in the flume-river.

With regard to changes in slope, the profiles of Horse Creek in Fig. 7.2 show that the slope of each of the divided channels is three times as large as the slope of the undivided stream. Increased slope on a divided reach is even more marked on the profile of the Green river near Daniel, Wyoming, where there is nearly a sixfold increase in slope after the river divides (Fig. 7.2). In the example of the New Fork river (Fig. 7.9) the steepening is less obvious, primarily because there is a steep riffle between stations 200 and 500 which tends to increase the average slope of the upper 1,000 ft of the mapped reach. If, however, the divided reach between stations 1700 and 2400 is compared with the undivided reach from stations 1500 to 1700,

TABLE 7.1
Ratio of hydraulic factors of divided to undivided reaches of braided streams
(natural rivers and flume-river)

	Green river near Daniel, Wyo.	New Fork river near Pinedale, Wyo.		Flume at California Institute of Technology			
		Reach 1 (1953)	Reach 2 (1954)	Feb. 15 Stations 10 and 14	Feb. 16 Stations 14 and 22	Feb. 18 Stations 10 and 14	Mar. 5 Stations 12 and 38
Area	1·3	1·03	1·6	0·94	1·08	0·78	1·07
Width	1·56	1·83	2·0	1·05	1·34	1·48	1·70
Depth	0·88	0·56	0·79	0·90	0·80	0·52	0·63
Velocity	0·77	0·97	—	1·06	0·93	1·27	0·93
Slope	5·7	2·3	1·4	1·3	1·4	1·9	1·7
Darcy–Weisbach resistance factor	10·5	1·3	—	1·1	1·3	0·63	1·25

steepening of the slope in the undivided part is again apparent (Table 7.1).

As the table shows, there is more variation in the ratios in slopes than there is in the ratios of widths and of depths in divided and undivided reaches. This is due primarily to one example in a natural river, the Green river near Daniel, for which the slope of the divided channel was nearly six times that of the undivided one (Fig. 7.7).

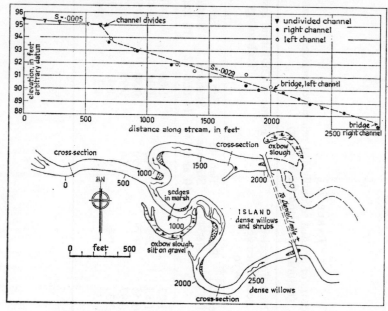

Fig. 7.7 Plan and profile of Green river near Daniel, Wyoming.

With the exception of this one large ratio, however, the natural rivers have ratios of 1·4 and 2·3, and the ratios in the flume ranged from 1·3 to 1·9.

The changes in width, depth and slope caused by island development in the flume-river are of the same order of magnitude as comparable changes in the natural rivers studied.

The change of cross-sectional area depends upon the relative amounts of change of width and depth. The division by an island caused the cross-sectional area to increase in all three reaches of the natural rivers studied, but in the flume there was an increase in cross-sectional area in two instances and a decrease in two instances. The velocity in the divided channels was somewhat less in the two natural rivers for which data are available, but division in the flume caused the velocity to increase in two reaches and to decrease in others. Non-uniform changes caused by island growth can also be seen in the Darcy–Weisbach resistance factor **f** (Table 7.1).

These data indicate that changes occur in nearly all the hydraulic parameters when a channel divides (for some discussion, see Franzius, 1936, and Chien, 1955). The adjustments are independent; it is difficult to specify in advance how an alteration of conditions will be taken up by the dependent factors.

Although island formation and changes in the channel are often too rapid for complete adjustment to equilibrium conditions, some reaches of braided rivers are very stable indeed. We infer, accordingly, that a braided pattern in common with other patterns can represent a condition of quasi-equilibrium.

STRAIGHT CHANNELS

In our experience truly straight channels are so rare among natural rivers as to be almost non-existent. Extremely short segments or reaches of the channel may be straight, but reaches which are straight for distances exceeding ten times the channel width are rare.

The wandering thalweg

Fig. 7.8 shows in plan and profile a reach of Valley Creek near Downingtown, Pennsylvania. For 500 ft this channel is straight in a reach where the alluvium of the valley is 30 ft thick. Its sinuosity (ratio of thalweg length to valley length) is practically 1·0. The thalweg, or line of maximum depth, is indicated by a dashed line in the upper portion of the diagram. Though the channel itself is straight, the thalweg wanders back and forth from positions near one bank and then the other. This is typical of a number of nearly straight reaches which we have studied.

Along with the wandering thalweg it is not uncommon to find deposits of mud adjacent to the banks of straight channels. These commonly occur in alternating (as opposed to opposite) positions. A similar observation has been made by Schaffernak (1950, p. 45). Straight channels thus appear to bear a remarkable resemblance to meandering channels. Quraishy (1944) observed that a series of alternating shoals formed in the straight flume channel of his experiment; these he referred to as 'skew shoals' (p. 36). Similarly Brooks (1955, pp. 668 ff.) called the condition in which low channel bars

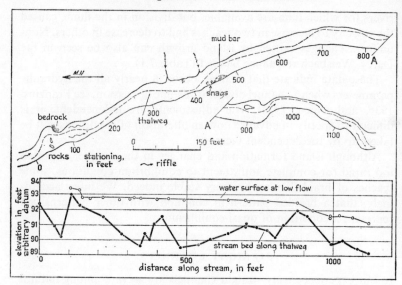

Fig. 7.8 Plan and profile of Valley Creek near Downingtown, Pennsylvania.

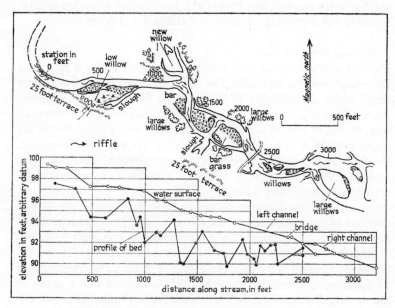

Fig. 7.9 Plan and profile of braided reach of New Fork Creek, near Pinedale, Wyoming.

formed alternately adjacent to the left and right walls of his straight flume a 'meander' condition.

Pools and riffles
Another characteristic of natural streams even in straight reaches is the occurrence of pools and riffles. This has been noted by Pettis (1927), Dittbrenner (1954) and Wolman (1955). Figs. 7.9, 7.10 and

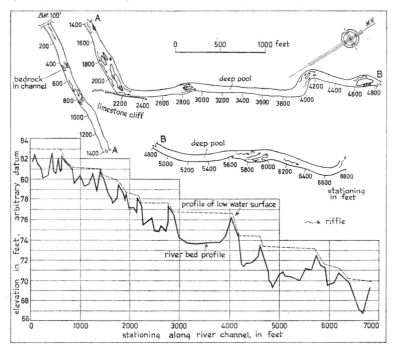

Fig. 7.10 *Plan and profile of Middle river, near Staunton, Virginia*

7.11 (the New Fork river near Pinedale, Wyoming, the Middle river near Staunton, Virginia, and the Popo Agie near Hudson, Wyoming) present respectively plans and profiles for a braided reach, a straight reach and a meandering reach. Profiles of these three examples are compared in Fig. 7.12, which reveals a similarity in the profiles of streams very dissimilar in pattern. That is, a straight channel implies neither a uniform stream bed nor a straight thalweg.

As demonstrated first by Inglis (1949, p. 147), the wave-length of a meander is proportional to the square root of the dominant discharge. One wave-length (a complete sine curve or 2π radians) encompasses twice the distance between successive points of inflection of the meander wave. It is well known that meandering channels characteristically are deep at the bend and shallow at the cross-over or point of inflection. Thus twice the distance between successive riffles in a straight reach appears analogous to the wave-length of a

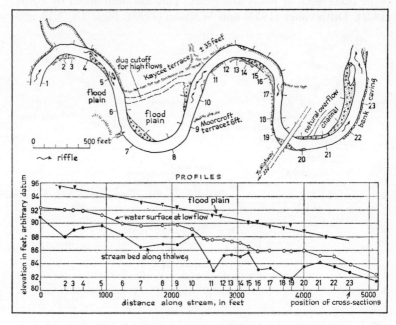

Fig. 7.11 Plan and profile of a meandering reach of the Popo Agie, near Hudson, Wyoming.

meander and should also be proportional to $q^{0.5}$. As an initial test of this hypothesis, bankfull discharge, which we consider equivalent to dominant discharge of the Indian literature, has been plotted in Fig. 7.13A against wave-length of meanders. The figure includes data from straight reaches for which wave-length is twice the distance between successive riffles.

The data in Fig. 7.13 include measurements of rivers in India from Inglis (1949), our own field measurements, and some flume data from Friedkin (1945) and Brooks (1955 and personal communication). The wave-lengths in Brooks's data obtained in a fixed-wall flume represent, as in the non-meandering natural channels, twice the distance between the riffles, or low bars, which he observed.

Fig. 7.13A demonstrates that, with considerable scatter, a relation between wave-length and discharge exists through a 10^8 variation in q. The data for straight channels do not vary from the average relation any more than those for meanders.

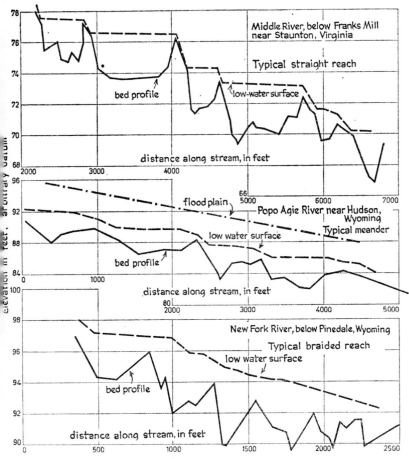

Fig. 7.12 Profiles of straight, meandering and braided reaches.

Inglis (1949, p. 144) had called attention to the fact that since width is also proportional to the square root of discharge, wave-length is a linear function of stream width. Inglis did not mention the fact, illustrated in Fig. 7.13B, that the scatter of points in the width–wave-length relation is less than that in the discharge-wave-length relation.

The relation in Fig. 7.13B is quite consistent, though it includes straight channels as well as meanders, and the widths range from less than 1 ft in the flume to 1 mile in the Mississippi river. The line drawn in Fig. 7.13B indicates that, in general, the ratio of wavelength to bankfull width varies from about 7 for small streams having

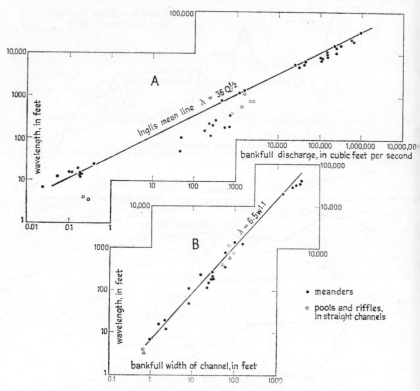

Fig. 7.13 Wavelengths of meanders and of riffles as functions of bankfull discharge and channel width.

widths of 1 to 10 ft, up to 15 for large rivers having widths in excess of 1,000 ft. Our data indicate that the relation is not a constant ratio but a power function having an exponent slightly larger than 1·0, specifically $\lambda = 6 \cdot 5 \omega^{1 \cdot 1}$.

Comparison of Figs. 7.13A and 7.13B leads us to postulate that the wave-length is more directly dependent on width than on discharge. It is argued later in this paper that in general, at a constant slope, channel width follows from discharge as a dependent variable.

We suggest, therefore, that wave-length is dependent on width and thus depends only indirectly on discharge. That this relation describes both the distance between riffles in straight channels and the wave-length of meanders leads us to conclude that the processes which may lead to meanders operate also in straight channels.

THE CONTINUUM OF CHANNELS OF DIFFERENT PATTERNS

The physical characteristics of the three specific channel patterns discussed in the preceding section suggest that all natural channel patterns intergrade. Braids and meanders actually represent extremes in an uninterrupted range of pattern. If we assume that the pattern of a stream is controlled by the mutual interaction of a number of variables, and the range of these variables in nature is continuous, then we should expect to find a complete range of channel patterns. A given reach of river may exhibit both braiding and meandering. In fact, Russell (1954) points out that the Meander river in Turkey, from which the very term *meandering* is derived, has some reaches which are braided and others which are straight.

This conception of transition in pattern, or interrelationship of channels of diverse pattern, is supported by the data in Fig. 7.14, where average channel slope is plotted as a function of discharge. Meandering, braided and straight channels are designated by different symbols. Reaches classed as meandering have a sinuosity of 1·5 or more; the value is arbitrary, but our experience suggests that, where it is equalled or exceeded, a stream would readily be accepted as meandering. The term *braid* is here applied to reaches where there are relatively stable alluvial islands, and in consequence two or more separate channels.

The data indicate that, in the rivers studies, braided and meandering channels are separated by a line described[1] by the equation

$$s = q\pi^{0·44} \tag{7.1}$$

For a given discharge, meanders occur on the smaller slopes. At the same slope a braided channel will have a higher discharge than a meandering one. Single straight channels – those with a sinuosity of less than 1·5 – occur throughout the range of slopes, giving support to the view that the separation of a true meander from a straight channel is arbitrary. Study of the sinuosities of these channels does not reveal an increasing sinuosity with decreasing slope.

[1] A list of symbols is given in the Appendix (p. 236).

In considering Fig. 7.14, it is important to remember that the data relate to natural channels, where specific variables are often associated. For instance, steep slopes are associated with coarse material. In a system which contains, as we have pointed out, a minimum of seven variables, a diagram such as Fig. 7.14 which treats only two of

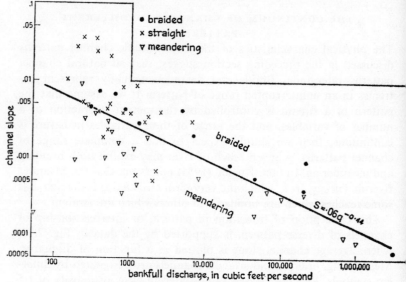

Fig. 7.14 *Slope and bankfull discharge: critical distinction of meandering from braiding.*

these cannot be expected to describe either the mechanism of adjustment or all theoretically possible conditions. Because it is drawn from nature, however, it does describe a set of conditions expectable in many natural channels.

Cottonwood Creek, near Daniel, Wyoming, strikingly illustrates an abrupt change from meandering to braiding (Fig. 7.15). The meandering reach has a slope of 0·0011, the braided reach one of 0·004. The difference in slope is accompanied by a change in median grain size from 0·049 ft in the meandering reach to 0·060 ft in the braided reach. It is clear that no change in discharge occurs along the length of the mapped channel, for no tributaries enter. The load carried is identical in the meandering and the braided reaches. Lack of evidence for rapid aggradation or degradation suggests that the meandering reach is in quasi-equilibrium; if this is true, then so is the braided reach. Braiding is here due not to excessive load, but appears to be a channel adjustment referable to the fortuitous occurrence in the alluvium of a patch of unusually coarse gravel.

In addition to differences in slope and discharge between meanders and braids, our observations indicate that in general, given identical slopes, braided channels have higher width/depth ratios than meandering channels.

The relation of channel pattern to slope and discharge is of particular interest in connection with the reconstruction of the alluvial

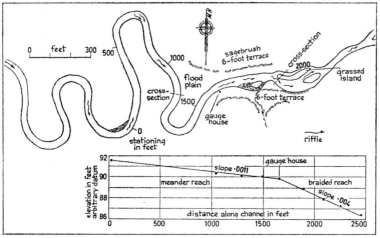

Fig. 7.15 *Change from meandering to braiding on Cottonwood Creek, near Daniel, Wyoming.*

landscape from profiles of terrace remnants and from alluvial fills. Changes of climate such as occurred in the Pleistocene might be accompanied by changes in precipitation and run-off. Resultant variations in stream flow might produce changes in the stream pattern without any accompanying change in the quantity or calibre of the load. An increase of flow could result in a meander becoming a braid; a decrease in discharge could make a braided channel a meandering one. Yet a change in the character of the load, such as an increase in calibre which the presence of a valley glacier might provide, could result in an increase in slope and a change from meandering to braiding without any change in the precipitation or discharge. The presence, then, of material of different sizes in successive valley fills suggests that at different periods in its history the river which occupied the valley may have had several distinctive patterns. As the pattern of major rivers in an area is a significant feature of the landscape, the application of such principles to historical geology may prove of some help in reconstructing the past.

THE NATURE OF CHANNEL ADJUSTMENT TO INDEPENDENT CONTROLS

Many geologists accept the idea that geology, climate and interrelated hydrological factors are the ultimate determinants of river morphology. Nevertheless the details of how control is exercised, particularly with respect to the hydraulic mechanisms, have not been fully described. This final section is meant to illustrate some parts of the process whereby the channel is ultimately controlled by geology and climate, through the effect of these factors on discharge and load.

Development of river width

It is reasonable to suppose that the quasi-equilibrium width of the channel is determined not by those floods which occupy the entire valley, but rather by discharges which attain or just overtop the banks of the channel (Wolman and Leopold, 1957). If the width is larger than necessary for quasi-equilibrium, the unused parts of the wide channel are taken over by vegetation which tends to stabilise the places where the roots are present and to induce deposition. The establishment of vegetation in unused parts of a channel provides a slow but effective way of reducing a width made excessive during high flows.

We have observed both in natural channels and in the flume that the wandering thalweg of a straight reach provides an additional mechanism for reducing channel width, by virtue of deposition on the insides of thalweg bends.

The width of a river is subject to constant readjustment if the banks are not well stabilised by vegetation. The magnitude of the readjustment depends on the nature of the banks and the amount and type of vegetation they support. In the eastern United States river banks generally tend to be composed of fine-grained cohesive material, and large trees typically grow out from the bank and lean over the stream. Their roots are powerful binding agents. In these conditions, width adjustments are small and slow. Only the large floods are capable of tearing out the banks. Width adjustments in the semi-arid West appear to be greater, for the material of the banks is more friable and the vegetation is less dense.

It might be supposed that if rivers from a great diversity of geographical areas were considered, flood discharge would correlate more closely with river width than it would with any of the other channel factors such as depth, velocity, slope or grain size, because the latter are apparently less directly controlled by discharge. Data from such diverse rivers as the arroyos in New Mexico, Brandywine Creek, Pennsylvania, the Yellowstone and Bighorn rivers, and the upper Green river indicate that such is the case. There is, of course, considerable variation in width of streams having equal discharges of a similar frequency. Nevertheless the variation among streams at a particular discharge is small relative to the change in width with increasing discharge in the downstream direction. The difference in width between streams having equal discharge appears to be related to sediment concentration and to the composition of the bed and banks.

The close relation between channel width and discharge is also shown by several runs in the flume (Table 7.2). Runs 23B and 30D each had a discharge of 0·033 cu. ft per sec. Despite the fact that the

TABLE 7.2

Comparison of channel factors in two flume runs at equal discharge
(California Institute of Technology flume, 1954)

	Run	
	23B	30D
Discharge (cu. ft per sec.)	0·033	0·033
Median grain size (ft)	0·00059	0·0036
Load (lb. per sec.)	0·00121	0·00202
Width (ft)	0·66	0·67
Hydraulic radius (ft)	0·0555	0·0320
Velocity (ft per sec.)	0·68	1·36
Slope	0·00175	0·0117
Darcy–Weisbach resistance factor	0·0575	0·0523

bed material in run 30D was six times as coarse as that in run 23B, and the slope about ten times as steep, each run developed the same width.

The rapidity with which the width of the flume channel adjusted to discharge indicates the close association of the two characteristics. More than twenty runs were made in which adjustment of width by bank erosion, and consequent change in mean depth, took place on the average is less than five minutes. After the initial adjustment no further change in width occurred during the remainder of a run. When the initial channel was wider than necessary for quasi-equilibrium, there was little or no adjustment to a narrower width except gradually in connection with a tendency of general aggradation or degradation.

We tentatively conclude that width is primarily a function of discharge. The magnitude of the effective discharge to which the width is adjusted is considered in another paper (Wolman and Leopold, 1957) where we argue that it occurs about once a year.

Channel roughness and resistance

One of the ways by which the lithological character of a basin affects river morphology is the size of the particles contributed to the debris load. To relate this effect to the hydraulic mechanisms by which the channel is adjusted, it is necessary to consider the interaction of grain size with other hydraulic factors.

Engineers are familiar with the concept (Rouse, 1950) that at high Reynolds numbers the Darcy–Weisbach resistance coefficient,

$$f \alpha \frac{gRs}{v^2} \tag{7.2}$$

is a function of relative roughness. Where the roughness is controlled by the grain size, relative roughness may be defined as the ratio of grain size to the depth of flow.

Empirical relations have been obtained, for flow in pipes, between resistance coefficient and relative roughness. In these relations the grain-size term has usually been defined on the basis of a uniform sand size. The magnitudes of roughness elements of pipes of various materials are often expressed in terms of equivalent grain size, i.e. uniform grains which give comparable resistance.

The beds of natural rivers, however, are often typified by a wide range of particle size. Because of variation of grain-size distribution over the channel bed, the sampling problem is important. Furthermore, even when an adequate sampling procedure is adopted, one must choose some representative size or some characteristic of the size distribution to typify the bed material.

Thus far no completely satisfactory method has been developed for relating grain size in natural rivers to resistance. Our approach to the problem has been an empirical one. The size distribution of grains making up the beds of several river reaches was measured, using a method described by Wolman (1954).

In our attempt to relate grain size to resistance, the grain-size parameter chosen is D_{84}; that is, 84 per cent of the material on the cumulative curve is finer than the size D_{84}. The 84 per cent figure is one standard deviation larger than the median size, a choice guided by our experience that this size gives the best correlation with resistance.

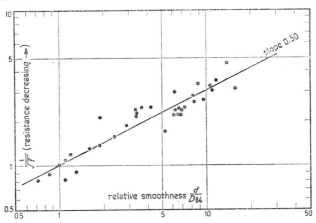

Fig. 7.16 Relation of relative smoothness to a resistance factor, for Brandywine Creek, Pennsylvania.

Fig. 7.16 shows an empirical relation between a resistance parameter $\frac{1}{\sqrt{f}}$ (where **f** is the Darcy–Weisbach coefficient) and relative smoothness (the ratio of mean depth of water, d, to grain size D_{74}). The data are for a number of reaches of Brandywine Creek, Pennsylvania (data from Wolman, 1955). The equation for the straight line drawn through the points is

$$\frac{1}{\sqrt{\mathbf{f}}} = \left(\frac{d}{D_{84}}\right)^{0.5} \tag{7.3}$$

The data from Brandywine Creek include measurements of the slope of the water surface which, in conjunction with the measurements of the bed material size, are available for very few locations elsewhere. To be correct, the slope used in the computation of **f** should be the slope of the energy grade line. Our data indicate, however, that for the purposes of this generalised analysis, the difference between the water surface and energy slopes is not significant.

For streams other than Brandywine Creek, our data do not include water surface slope but only the mean slope of the channel bed. The latter is usually a rough approximation to water surface slope, but its use constitutes an additional source of variance. It is not astonishing, therefore, that when such data are added to those for Brandywine Creek, as has been done in Fig. 7.17, the scatter is greater than

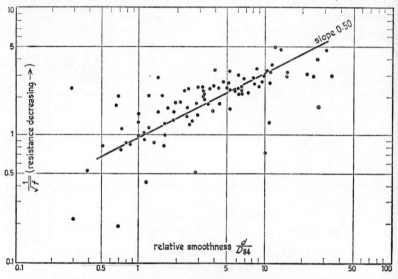

Fig. 7.17 Relation of relative smoothness to a resistance factor, for all data collected by the authors.

in Fig. 7.16. Nevertheless the mass of points encompasses the points for Brandywine Creek alone, and the two diagrams indicate in a general way how geology, through its effect on grain size, may influence the properties of river channels.

The computed resistance is probably not solely a function of the size of the bed material. Although the data largely represent streams with coarse beds in which the ratio of width to depth is usually greater than 20 to 1, the resistance in some of them may be affected by vegetation, channel alignment or other factors not attributable to grain roughness. If grain size is not the dominant factor controlling the resistance, the relation shown in Figs. 7.16 and 7.17 would not apply.

Resistance can also be studied in relation to bed configuration. In many natural channels with beds of fine material, the roughness of the channel is related to the dunes and ripples which form on the bed, rather than to the size of the discrete particles themselves. In the flume experiment dunes were characteristic of all runs in which the bed was composed of fine sand, D_{50} equal to 0·00059 ft. As Table 7.2 shows, in run 30D the grain size was six times as large as in run 23B. When each was experiencing the same discharge, however, the resistance factor in run 23B was approximately equal to that in run 30D. Thus the adjustments in velocity, depth and slope which accompanied the changes in the configuration of the bed of fine sand resulted in a resistance in the bed of fine material equivalent to that in the much coarser channel. Where bed configuration is important, the channel roughness can no longer be considered even partially independent. Roughness is controlled by changes which take place in velocity, depth and slope in response to changes in discharge and load. Detailed studies of sediment transport by Brooks (1955) support this conclusion.

Sediment transport, shear and resistance

The complexity of the problem indicated by these brief observations on bed configuration and resistance leads directly to a consideration of some illustrations of the interrelation of resistance, shear and sediment transport.

In some of the runs made in the flume, a deep trapezoidal channel (template E, run No. 20A, B, C) was moulded in the sand. Though the discharge nearly filled the channel, the sand making up the bed of the flume was not moving. Load fed at the upper end of the flume tended to move downstream as a front, sediment deposited in a given reach building up the bed and gradually reducing the depth and increasing the velocity in the reach. When the depth was decreased sufficiently, the grains passed over the new deposit and were carried downstream where they rolled over the end of the new deposit like foreset beds being deposited in a delta.

In run 20 (Table 7.3), stations 9 and 13 are upstream from the depositional front, and station 26 is downstream from the front. The data show that passage of the front through a given reach was accompanied by an increase in velocity and slope and by a decrease in hydraulic radius. The shear (computed as γRs) at station 26, in which no sediment was moving, very nearly equalled the shear at station 13 through which sediment was moving. The movement of sediment at station 13 was, however, associated with a much lower resistance.

At the outset of run 17 no sediment was being removed from the bed of the channel until sand introduced at the head of the flume moved down the channel. As the run progressed, the sediment load introduced into the flow moved progressively downstream as a thin sheet of material along the bed. After passage of this sheet, movement on the bed was continuous. Vertical velocity profiles taken

TABLE 7.3

Comparison of reaches of channel before and after passage of sediment front

(California Institute of Technology flume, 18 March 1954)

	Run		
	20A	20B	20C
Station	9	13	26
Time	8.00 a.m.	3.50 p.m.	3.50 p.m.
Position of front relative to the station	1 ft downstream	5 ft downstream	8 ft upstream
Discharge (cu. ft per sec.)	0·086	0·086	0·086
Load (lb. per sec.)	0·00253	0·00253	No movement
Velocity (ft per sec.)	1·18	1·18	0·66
Hydraulic radius (ft)	0·064	0·062	0·120
Slope	0·0068	0·0041	0·0018
Shear ($\tau_0 = \gamma Rs$)	0·027	0·016	0·014
Darcy–Weisbach resistance factor	0·080	0·047	0·130

at station 36·8 before and after passage of the sheet are shown in Fig. 7.18. The grain size remained constant and the rate of change of velocity with depth was greater when sediment was not moving on the bed. As the rate of change of velocity with depth is directly proportional to the shear velocity and the Von Karman kappa, it is tentatively assumed that the transport of sediment alters the value of kappa as found by Vanoni (1946).

A series of runs in the flume was designed to show the effect of various rates of introduction of sediment load under constant conditions of discharge, flume slope and template size. The initial channel was moulded with a trapezoidal section having a top width of 9 in. and a depth of 2½ in. Experience had shown that a discharge of 0·033 cu. ft per sec. would require no readjustment of this width.

The flume slope was set at 0·0103. In the first run in the series no load was introduced. There was a gradual winnowing of the bed during which there was active movement over the central 3½ in. of the total bed width of 7 in. After eighteen hours of flow with no load being introduced, the whole channel became winnowed of finest grains and further degradation had in effect ceased.

During this degradation the slope flattened slightly and the bed coarsened until little or no movement occurred. At this discharge

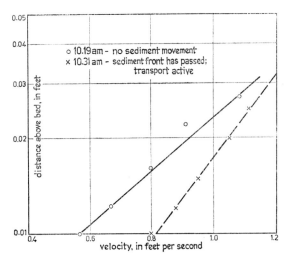

Fig. 7.18 Vertical profiles of velocity in the flume; at a given grain size, shear velocity is not the role determinant of bed transport.

the stream became unable to move the material on its bed; thus this discharge was analogous to low flow in a natural river with a coarse bed. Because the discharge was too low to be effective, the slope could not be altered.

Runs were made at the same discharge and initial slope introducing load at the rate of 24 gm. per minute and 55 gm. per minute. In each run there was some steepening of the reach farthest upstream as in previous instances in which a braided channel developed. The principal part of the flume, however, through which moved the bulk of sediment minus part of the coarse fraction deposited in the developing island, remained relatively stable. The slope of this lower reach became 0·0117 in the run (No. 30D) during which 55 gm. per minute was introduced, and 0·0112 in the run (No. 31) of 24 gm. per minute.

The most striking difference between these two runs (Nos. 30D and 31) in which different quantities of load were introduced, was the width of the band of moving sediment in the long reach below the developing braid. The larger introduced load (run 30D) was associated with movement of the load over most of the bed width. With the smaller introduced load the movement on the bed was confined to a narrow band in the centre of the channel.

This same effect was shown by another series of runs in which all particles coarser than 3·32 mm. had been sieved out of the introduced sand load. The width of the moving band of sediment appeared to depend on the rate of introduction of load, as shown in Table 7.4.

TABLE 7.4

Relation of width of moving band of sediment to rate of introduction of load

(Measured at station 22, California Institute of Technology flume, 1954)

Run	Rate of load introduction (gm. per minute)	Width of band of moving bed sediment (in.)
32	56	$2\frac{1}{4}$
33	115	$3\frac{1}{2}$
34	175	$4\frac{1}{2}$

In summary, the bed sediment movement in the flume tended to concentrate in the centre of the bed, not only when the bed was flat, but even after a central ridge began to develop. When no central ridge was being built, the width of the band of moving sediment increased with the amount of load coming into the reach.

Response of slope to changes of load in the flume

Another series of runs was designed to test the relation between sediment load and slope at a constant discharge. The cross-section remained practically constant during the series. The data (Table 7.5) were obtained under constant initial conditions except rate of introduction of load. All material coarser than 3·32 mm. had been screened from the introduced load, and as a result no island or bar appeared. When the load was introduced at various rates at the same discharge (runs 32 to 35), the slope became adjusted to nearly the same value whether the reach was stable or aggrading. Aggradation was characterised by a gradual and uniform rise of the bed all along the flume.

Despite a fourfold increase in load, slope remained essentially constant. Although this observation appears to conflict with Gilbert's conclusion (Gilbert, 1914), analysis of his data shows that the

TABLE 7.5

Effect of changing load on stream slope, all other conditions being kept constant

(California Institute of Technology flume, 1954)

Run	Discharge (cu. ft per sec.)	Load introduced (gm. per minute)	Slope	Condition of channel
32	0·033	56	0·0105	Nearly stable
33	0·033	115	0·0105	Slow aggradation
34	0·033	175	0·0109	Aggradation
35	0·033	212	0·0110	Rapid aggradation

increase of slope which took place with increasing load in his experiments was also accompanied by an increase in roughness. In our experiment there was but a relatively small change in either slope or resistance.

Though slope was little affected by changes of total load in the flume experiments, there was some indication that at constant grain size, relatively steep slope is associated with small discharge and flat slope with large discharge. The data do not permit a definite statement on this relation. Considering the downstream rate of change of velocity, depth and width with discharge in an average river (Leopold and Maddock, 1953, p. 26) under the assumption of constant roughness (f), slope must decrease with discharge as

$$s \propto q^{-0.20} \tag{7.4}$$

The part which changing cross-sectional area downstream plays in this relation is unknown. Tentatively, it is believed that at constant grain size, river sections having different effective discharges should differ in slope, larger discharges being associated with flatter slopes.

OBSERVATIONS OF ADJUSTMENTS IN NATURAL CHANNELS

Measurements in the flume are complemented by examples drawn from larger rivers illustrating the same principles of adjustment.

Table 7.6 indicates the nature of the adjustments which accompany changes in sediment load at constant discharge. Unfortunately there are few rivers for which data on the flow parameters, including slope, are available, and the data for these few suffer because suspended load but not bed load was measured.

TABLE 7.6

Examples from river data of adjustments of depth, velocity and slope to changes in suspended load at constant discharge and width

Stage	Elkhorn river near Waterloo, Nebraska, 26 March to 9 April 1952		Section A-2, Rio Grande at Bernalillo, New Mexico, April to June 1952	
	Rising	Falling	Rising	Falling
Discharge (cu. ft per sec.)	6,000	6,000	4,700	4,700
Width (ft)	255	263	275	275
Velocity (ft per sec.)	4·90	4·56	5·5	5·0
Depth (ft)	4·8	5·0	3·1	3·4
Slope	0·00038	0·00046	0·00094	0·00084
Suspended load (tons per day)	96,000	74,000	46,000	21,000
Elevation of bed above arbitrary datum (ft)	7·8	7·8	3·2	3·2
Shear ($\tau_0 = \gamma ds$) (lb. per sq. ft)	0·114	0·143	0·182	0·178
Resistance factor $\left(f = \dfrac{8\,gds}{v^2}\right)$	0·019	0·028	0·025	0·029
Grain size, suspended material D_{50} (mm.)	0·03	0·04	0·13	0·13
Grain size, bed material D_{50} (mm.)			0·32	0·32

In Table 7.6 the flow characteristics of the Elkhorn river, which are compared at a discharge of 6,000 cu. ft per sec., represent the rising and falling sides of a hydrograph. The rising stage was typified by a lesser depth and lesser slope, and a greater velocity and greater load than the falling stage. The elevation of the bed remained the same. The larger load of the rising stage was associated with lesser roughness. The lower resistance led to a greater velocity and a lesser depth despite the slightly lesser slope. The product of depth times slope, which is directly proportional to the shear, was less in the rising stage, and thus the larger load was associated with the lesser shear. It is presumed that the increase in velocity associated with the lesser roughness was of greater importance than the relatively small difference in shear in promoting equilibrium with the large load.

The data on the Rio Grande at Bernalillo (Table 7.6) show another type of adjustment. The rising stage was characterised by a lesser depth, greater velocity and greater load than the falling stage. In the falling stage slope was somewhat less, and thus the smaller load was associated with the lesser shear. Here, then, the greater shear was associated with the larger load, but, as in the Elkhorn river, the total resistance was less when the load was large.

SUMMARY AND INTERPRETATION

From these diverse observations on the nature of channel adjustments to independent controls we draw several tentative conclusions:

1. Channel width is largely determined by discharge; the effective discharge we believe to be that corresponding approximately to bankfull stage.
2. Where resistance is controlled by the roughness provided by discrete particles on the bed, the resistance factor may be correlated with particle size.
3. Where the configuration of the bed enters into the determination of roughness, the resistance is a function of the discharge and load and represents a simultaneous adjustment of velocity, depth and slope.
4. The load transported at a given discharge is not solely a function of shear and grain size. At a constant shear, decreased resistance appears to be associated with greater transport. The resistance is not necessarily governed by calibre of load but may be determined principally by bed configuration.
5. In the flume-river the width of the band of moving sediment was often less than the width of the channel; it was related to the quantity of load being transported. If such a band exists in a natural channel, errors might result from estimates of total load based on computations of load per foot of width multiplied by the total width.
6. Where no change in roughness occurs at constant discharge, a large increase in load is not accompanied by an appreciable increase in slope. Aggradation and degradation may alike occur, therefore, without change in slope.

We shall now attempt to construct a general though oversimplified picture of the way in which the shape and pattern of a natural channel may be determined by the river itself within the framework provided by climate, rocks and physiography.

The magnitude and character of run-off are determined by climate and lithological factors which are quite independent of the channel system. The topographical and lithological character of the catchment helps to determine not only the characteristics of the run-off but also the load of debris delivered to the channel system. Quantity and nature of load are greatly influenced and are often primarily governed by geological factors.

Exceptional run-off is accompanied by conditions which increase the movement of soil and rock debris. Of particular importance are relatively large amounts of rainfall, often at high intensity, and saturated soils. Moreover large volumes of run-off are associated with relatively large depths of flow both in rills and in overland flow. These depths are accompanied by relatively high flow velocity and increased transport of debris to the channel system. For such reasons large flows are usually accompanied by large loads.

In a given period of time there are far fewer large flows than small or moderate ones. An intermediate range of discharge includes flows that occur often enough in time and possess sufficient vigour to constitute the effective discharge (Wolman, 1956). During these flows the river can move the material on its bed and in its banks and thus is capable of modifying its shape and pattern.

Channel width is primarily determined by discharge. Widening is rapid relative to other changes and the channel generally tends to become adjusted at the width provided by large flows near the bankfull stage.

The roughness of any reach of a channel is governed, initially at least, by the geological character of the bed material supplied by the drainage basin. Although abrasion and sorting may modify the material on the bed, the rock character partly determines the extent of such modification. The roughness may be primarily grain roughness. Alternatively, owing to the size distribution of the particles and their movement, the rugosity may be due to the configuration of the dunes, ripples or waves in which the particles on the bed arrange themselves.

The roughness of the boundary affects the stress structure and the velocity distribution in the flowing water. In addition, turbulent eddies near the bed produced by the roughness elements lift particles off the bed. These alternately move forward in the current and settle back to the bed under the influence of gravity. Not only the resistance then, but also the movement of debris through turbulence and shear, is related to roughness.

At a given discharge, with width fixed thereby, velocity, depth and slope become mutually adjusted. To visualise the nature of this adjustment, one may assume temporarily a value of slope. With slope given, velocity and depth must mutually adjust to meet two requirements. The first is that the product of width, depth and velocity is equal to the value of discharge, or

$$q = wdv \tag{7.5}$$

The second is the relation of resistance to depth, slope and velocity, which is expressed by the equation for the Darcy–Weisbach coefficient,

$$\mathbf{f} \propto \frac{gRs}{v^2} \tag{7.2}$$

The resistance f can be assumed to be fixed by the materials in the bank and on the bed.

These two equations must be satisfied. They contain six factors, but where four of the factors are fixed or temporarily assumed, these two equations fix the remaining two variables.

At a given discharge, width is considered fixed. Resistance is determined by the nature of the material in bed and banks, channel configuration and bed condition. Slope has been temporarily assumed. Hence velocity and depth become determined by the two equations.

If this combination of depth, velocity and slope constitute a channel that will transmit the given discharge and load without erosion or deposition, the slope is not altered. If the hydraulic factors of the channel are not in equilibrium with the imposed load, erosion or deposition will occur. By such action the slope of relatively long reaches of the channel is altered, and, as the alteration occurs, the velocity and depth are also changed.

This explanation provides a perspective upon the interaction of the several variables in producing the channels observed in nature. The variables are interdependent, but their relative order of importance is believed to be generally as outlined. If correct reasoning is to be undertaken about the relations of geological and other factors to fluvial processes, it is necessary to determine the relative independence of the variables concerned.

We reason further that if, for example, a river is able to move the size of material in the load but is unable to transport the total quantity of load delivered to a given reach, deposition may occur in that reach with little or no change of slope. If deposition occurs on the bed of an alluvial river during a flood, this deposition at constant width is customarily accompanied by a decrease in depth and consequent increase in velocity through reduction in cross-sectional area. The increase in velocity at the expense of depth implies that deposition on the bed does not cause an equal rise in elevation of the water surface and of the bed.

Our studies indicate that an increase in load at constant discharge is usually accompanied by a decrease in bed roughness which serves to increase the velocity. The adjustment in depth and the smoothing of the bed are inextricably related. It is of course possible that the increase in capacity resulting from the increase in velocity would still be inadequate to transport the total load, in which event both the bed and the water surface would rise.

If a stream can move only some of the sizes in the load provided to the reach, the process of adjustment is similar to the one described above, save for the fact that, by the winnowing from the bed of certain-sized fractions only, the mean size of bed material is increased. The increased size will be associated with an increased slope over the entire reach, increasing the total capacity for transport of all movable sizes.

APPLICATION OF OBSERVATIONS ON CHANNEL ADJUSTMENT TO THE PROBLEM OF CHANNEL PATTERN

As explained, eight and possibly more variables enter in a consideration of natural stream channels: discharge, amount of sediment load, calibre of load, width, depth, velocity, slope and roughness. Each of these factors varies as a continuous function; that is, within the limits of observed values, any intermediate value is possible. The factor having the largest range of values is discharge, for in natural channels it can vary from zero to any amount less than 10 millon cu. ft per sec. Load, expressed as concentration by weight, varies from nearly 0 to about 500,000 p.p.m. Calibre of load, expressed as median grain size, may vary from a value near zero to one of 10 ft. Width of natural channels varies between a few tenths of a foot to about 20,000 ft, mean depth from zero to an amount less than 80 ft, mean velocity from near zero to less than 25 ft per sec., slope from near zero to 1·0, and resistance, expressed as the dimensionless Darcy–Weisbach number, from about 0·001 to about 0·30.

Combinations observed in nature are far more restricted than the possible permutations of the values of the eight variables. To our knowledge, for example, rivers capable of discharging more than

1 million cu. ft per sec. have slopes not in excess of 0·0009 and usually less than 0·0002.

Braided, meandering and straight channel patterns all occur in nature throughout the whole range of possible discharges. However, our observations indicate that braids tend to occur in channels having certain combinations of values of the flow factors, and that meanders occur in different combinations. The straight pattern can occur in either kind of combination. Specifically, at a given discharge, braids seldom occur in channels having slopes less than a certain value, while meanders seldom occur at slopes greater than that same value.

These considerations, together with the details discussed earlier in this study, lead to one of the principal points of the present discussion: that there is a continuum of natural stream channels having different characteristics that are reflected in combinations of values of the hydraulic factors. Each of the channel patterns, braided, meandering and straight, is associated with certain of these combinations. The combinations of hydraulic factors that exist in most natural channels are those which represent quasi-equilibrium between the independent factors of discharge and amount of calibre of load on the one hand, and the dependent factors of form and hydraulic characteristics of the channel on the other. Braided, meandering and straight patterns, in this conception, are all among the forms of channels in quasi-equilibrium.

This conception fits the observations that a given channel can change in a short distance from a braid to a meander or vice versa, that the divided channels of a braid may meander, and that a meandering tributary may join a braided master stream. Such changes in a given channel or such different channels in juxtaposition can be attributed to variations in locally independent factors.

REFERENCES

BROOKS, N. H. (1955) 'Mechanics of streams with movable beds of fine sand', *Proc. Amer. Soc. Civ. Eng.*, LXXXI 668(1)–668(28).
CHIEN, NING (1955) 'Graphic design of alluvial channels', ibid., LXXXI 611(1)–611(17).
DIETZ, R. A. (1952) 'The evolution of a gravel bar', *Mo. Bot. Garden Ann.*, XXXIX 249–54.
DITTBRENNER, E. F. (1954) 'Discussion of river-bed scour during floods', *Proc. Amer. Soc. Civ. Eng.*, LXXX 479(13)–479(16) (separate publ. no. 479).
EINSTEIN, H. A. (1950) *The Bed-load Function for Sediment Transportation in Open Channel Flows*, U.S. Dept. Agric. Tech. Bull. 1026, 71 pp.

FRANZIUS, O. (1936) *Waterway Engineering* (Cambridge, Mass., Inst. Technology Press) 527 pp.

FRIEDKIN, J. F. (1945) *A Laboratory Study of the Meandering of Alluvial Rivers* (Vicksburg, Miss., U.S. Waterways Expt. Sta.) 40 pp.

GILBERT, G. K. (1914) *The Transportation of Debris by Running water*, U.S. Geol. Survey, Prof. Paper 86.

INGLIS, C. C. (1940) *Central Board Irrigation Annual Rept. 1939–40*, Publ. no. 24.

—— (1949) *The Behaviour and Control of Rivers and Canals* (Poona, Central Waterpower Irrigation and Navigation Research Sta.), XIII (1) 283 pp.

JACKSON, J. R. (1834) 'Hints on the subject of geographical arrangement and nomenclature', *J. Roy. Geogr. Soc.*, IV 72–88.

LEOPOLD, L. B., and MADDOCK, T., JR. (1953) *Hydraulic Geometry of Stream Channels and some Physiographic Implications*, U.S. Geol. Survey, Prof. Paper 252, 57 pp.

——, and MILLER, J. P. (1956) *Ephemeral Streams: Hydraulic Factors and their Relation to the Drainage Net*, U.S. Geol. Survey, Prof. Paper 282-A.

PEALE, A. C. (1879) 'Report on the geology of the Green River district', in Hayden, F. V., *U.S. Geol. and Geog. Survey Terr. 9th Ann. Rept.*, 720 pp.

PETTIS, C. R. (1927) *A New Theory of River Flood Flow* (privately printed) 68 pp.

QURAISHY, M. S. (1944) 'The origin of curves in rivers', *Current Sci.*, XIII 36–9.

ROUSE, H. (ed.) (1950) *Engineering Hydraulics* (New York, John Wiley & Sons) 1013 pp.

RUBEY, W. W. (1952) *Geology and Mineral Resources of the Hardin and Brussels Quadrangles* (Illinois), U.S. Geol. Survey, Prof. Paper 218, 179 pp.

RUSSELL, R. J. (1954) 'Alluvial morphology of Anatolian rivers', *Ann. Assoc. Amer. Geogr.*, XLIV, no. 4, 363–91.

SCHAFFERNAK, F. (1950) *Grundriss der Flussmorphologie und des Flussbaues* (Wien, Springer-Verlag) 115 pp.

VANONI, V. A. (1946) 'Transportation of suspended sediment by water', *Trans. Amer. Soc. Civ. Eng.*, no. 111, 67–133.

WOLMAN, M. G. (1954) 'A method of sampling coarse river-bed material', *Trans. Amer. Geophys. Union*, XXXV, no. 6, 951–6.

—— (1955) *The Natural Channel of Brandywine Creek, Pa.*, U.S. Geol. Survey, Prof. Paper 271.

——, and LEOPOLD, L. B. (1957) *River Flood Plains: Some Observations on their Formation*, U.S. Geol. Survey, Prof. Paper 282-C (in press).

APPENDIX: SYMBOLS

d mean depth, defined as ratio of cross-sectional area to width

D_{50} median size of sediment particle; subscripts 25, 75 or 84 refer to percentage of sample finer than specified size

f Darcy–Weisbach resistance factor

g acceleration due to gravity

q water discharge (cu. ft per sec.)

R hydraulic radius (ft)

s slope of water surface

v   mean velocity, defined as quotient of discharge divided by cross-sectional area
w   width
γ specific weight of water, or 62·4 lb. per cu. ft
T_o intensity of boundary shear

8 River Meanders and the Theory of Minimum Variance

WALTER B. LANGBEIN and
LUNA B. LEOPOLD

So ubiquitous are curves in rivers and so common are smooth and regular meander forms that they have attracted the interest of investigators from many disciplines. Also, investigations of the physical characteristics of glaciers and oceans have led to the recognition that analogous forms occur in melt-water channels developed on glaciers and even in the currents of the Gulf Stream. The striking similarity in physical form of curves in these various settings is the result of certain geometric proportions apparently common to all, that is, a nearly constant ratio of radius of curvature to meander length and of radius of curvature to channel width (Leopold and Wolman, 1960, p. 774). This leads to visual similarity regardless of scale.

Explanations offered for river meanders include such processes as

1. Regular erosion and deposition.
2. Secondary circulation in the cross-sectional plane.
3. Seiche effect analogous to lake seiches.

Attempts, however, to utilise these theories to calculate the forms of meanders have failed.

Although various phenomena – including some of those mentioned above, such as cross-circulation – are intimately involved in the deviation of rivers from a straight course, the development of meanders is patently related to the superposition of many diverse effects. Although each individual effect is in itself completely deterministic, so many of them occur that they cannot be followed in detail. As postulated in this paper, such effects can be treated as if they were stochastic – that is, as if they occurred in random fashion.

This paper examines the consequences of this postulate in relation to (1) the planimetric geometry of meanders, and (2) the variations in such hydraulic properties as depth, velocity and slope in meanders as contrasted with straight reaches.

The second problem required new data. These were obtained during the snow-melt season in 1959–64.

MEANDER GEOMETRY

Meandering in rivers can be considered in two contexts. The first involves the whole profile from any headwater tributary downstream through the main trunk – that is, the longitudinal profile of the river system. The second includes a meandering reach in which the channel in its lateral migrations may take various planimetric forms or paths between two points.

The river network as a whole is an open system tending towards a steady state, in which several hydraulically related factors are mutually interacting and adjusting – specifically: velocity, depth, width, hydraulic resistance and slope. These adjust to accommodate the discharge and load contributed by the drainage basin. The adjustment takes place not only by erosion and deposition but also by variation in bed form affecting hydraulic resistance and thus local competence to carry debris. Previous application of the theory of minimum variance to the whole river system shows that downstream change of slope (long profile) is intimately related to downstream changes in other hydraulic variables. Because discharge increases in the downstream direction, minimum variance of power expenditure per unit length leads towards greater concavity in the longitudinal profile. However, greater concavity is opposed by the tendency towards uniform power expenditures on each unit of bed area.

In the context of the whole river system, a meandering segment tends to provide greater concavity by lengthening the downstream portion of the profile. On account of the increase in the concavity of profile, the product of discharge and slope (power per unit length) becomes more uniform along a stream that increases in flow downstream. Thus the meander decreases the variance of power per unit length. Though the occurrence of meanders affects the stream length and thus the river profile, channel slope in the context of the whole river is one of the adjustable and dependent parameters, being determined by mutual interaction with the other dependent factors.

In the second context of any reach or segment of river length, average slope is given, and local changes of channel plan must maintain it. Between any two points on the valley floor, however, a variety of paths are possible, any of which would maintain the same slope and thus the same length. A thesis of the present paper is that the meander form occupies a most probable among all possible paths. This path is defined by a random-walk model.

Path of greatest probability between two points

The model is a simple one; a river has a finite probability p[1] to deviate by an angle $d\varphi$ from its previous direction in progressing an elemental distance ds along its path. The probability distribution as a function of deviation angle may be assumed to be normal (Gaussian), which would be described by

$$dp = c\ \exp{\tfrac{1}{2}} \left(\frac{d\varphi/ds}{\sigma}\right)^2 \tag{8.1}$$

where σ is the standard deviation and where c is defined by the condition that $\int dp = 1\cdot 0$.

The actual meander path then corresponds to the most probable river path between two points A and B, if the direction of flow at the point A and the length of the path between A and B are fixed and if the probability of a change in direction per unit river length is given by the probability distribution above (cf. von Schelling, 1951, 1964). Von Schelling demonstrates that the arc length s is defined by the elliptic integral

$$s = \frac{1}{\sigma} \int \frac{d\varphi}{\sqrt{[2(\alpha - \cos\varphi)]}} \tag{8.2}$$

where φ is the direction angle measured from the line AB and α is a constant of integration. It is convenient to set

$$\alpha = \cos\omega \tag{8.3}$$

in which ω becomes the maximum angle the path makes from the origin with the mean direction. Curves for $\omega = 40°, 90°, 110°$, which all show patterns characteristically seen in river meanders, are given in Fig. 8.1. Von Schelling (1951) showed that a general condition for the most frequent path for a continuous curve of given length between the two points A and B is

$$\Sigma \frac{\Delta s}{\rho^2} = \text{a minimum} \tag{8.4}$$

where Δs is a unit distance along the path and ρ is the radius of curvature of the path in that unit distance. But since

$$\rho = \left(\frac{\Delta s}{\Delta \varphi}\right) \tag{8.5}$$

[1] A list of symbols is given in the Appendix (p. 262).

River Meanders and the Theory of Minimum Variance

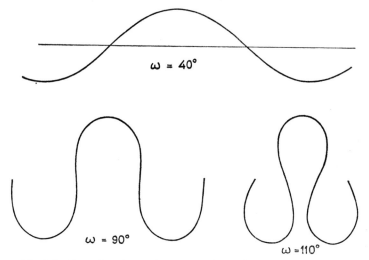

Fig. 8.1 Examples of most frequent random-walk paths of given length between two specified points on a plane (adapted from von Schelling).

where $\Delta\varphi$ is the angle by which direction is changed in distance Δs, therefore,

$$\Sigma \frac{(\Delta\varphi)^2}{\Delta s} = \text{a minimum.} \tag{8.6}$$

Since the sum of all the directional changes is zero, or

$$\Sigma \Delta\varphi = 0 \tag{8.7}$$

equation (8.6) is the most probable condition in which the variance (mean square of deviations in direction) is minimum.

The sine-generated curve

For graphing meanders it is easier to make use of the proposition that

$$\frac{d\varphi}{ds} = \sigma \sqrt{\left\{ 2\left(1 - \cos\omega\left[1 - \left(\frac{\varphi}{\omega}\right)^2\right]\right)\right\}} \tag{8.8A}$$

approximates closely to

$$\frac{d\varphi}{ds} \sigma \sqrt{[2(\cos\varphi - \cos\omega)]} \tag{8.8B}$$

where angles ω and φ are deviations from the central axis with the downstream direction as zero. With this simplification,

$$\varphi = \omega \sin \frac{\sigma \sqrt{[2(1-\cos\omega)]}}{\omega} s \tag{8.9}$$

or

$$\varphi = \omega \sin\frac{s}{M} 2\pi \tag{8.10}$$

where M is the total path distance along a meander, and where ω is the maximum angle the path makes from the mean down-valley direction. The maximum possible value of ω is 2·2 radians, for at this angle meander limbs intersect. When ω equals 2·2 radians, the pattern is a closed figure 8.

Comparison of equations (8.9) and (8.10) indicates that

$$M = \frac{2\pi}{\sigma} \frac{\omega}{\sqrt{[2(1-\cos\omega)]}} \tag{8.11}$$

all angles being in radians, as before.

Inasmuch as the ratio $\omega/\sqrt{2(1-\cos\omega)}$ is nearly constant ($=1\cdot05$), over the range of possible values of ω, meander path lengths, M, are inversely proportional to the standard deviation σ; thus, $M = 6\cdot6/\sigma$.

Equation (8.10) defines a curve in which the direction φ is a sine function of the distance s along the meander. Fig. 8.2 shows a theoretical meander developed from a sine function. The meander thus is not a sine curve, but will be referred to as a *sine-generated curve*.

Meander loops are thus generated by the sine curve defined by equation (8.10); the amount of horseshoe looping depends on the value of ω. This means that at relative distance $S/M = \frac{1}{2}$ and 1, φ has a value of zero; the channel is locally directed in the mean direction. At distance $S/M = \frac{1}{4}$ and $\frac{3}{4}$, the value of φ has its largest value ω, as can be seen in Fig. 8.2. The graph on Fig. 8.2A has been constructed for a value of $\omega = 110°$, and corresponds almost exactly to the 110° curve of Fig. 8.1 calculated by the exact equations of von Schelling.

The plan view of the channel is not sinusoidal; only the channel direction changes as sinusoidal function of distance (Fig. 8.2B). The meander itself is more rounded and has a relatively uniform radius of curvature in the bend portion. This can be noted in the fact that a sine curve has quite a straight segment as it crosses the x-axis.

The tangent to the sine function in Fig. 8.2B at any point is $\Delta\varphi/\Delta s$, which is the reciprocal of the local radius of curvature of the meander. The sine curve is nearly straight as it crosses the zero axis. Therefore, the radius of curvature is nearly constant in a meander

River Meanders and the Theory of Minimum Variance

bend over two portions covering fully a third of the length of each loop.

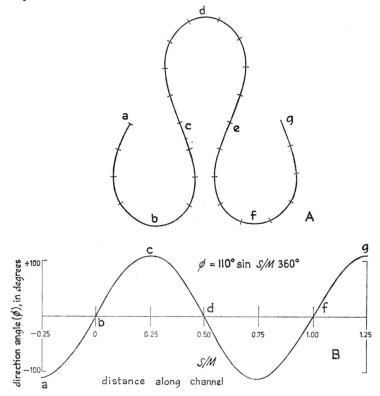

Fig. 8.2 A, *Theoretical meander in plan.* B, *Plot of direction angle φ against distance along channel path.*

Analysis of some field examples

Some field examples presented in Figs. 8.3A–E will now be discussed, with a view to demonstrating that among a variety of meander shapes the sine-generated curve fits the actual shape quite well and better than alternatives.

Fig. 8.3A, upper part, shows the Greenville bends at Greenville, Mississippi, before artificial cut-offs changed the pattern. The crosses in the lower diagram represent values of channel direction, φ – relative to a chosen zero azimuth, which for convenience is the mean downstream direction – plotted against channel distance. The full curve in the lower diagram is a sine curve chosen in wave-length

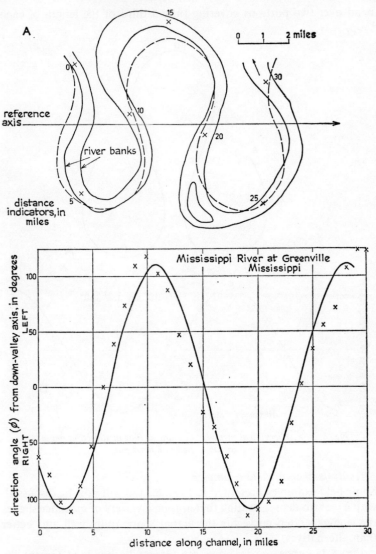

Fig. 8.3A Plan of meandering reach: sine-generated curve and plots of channel direction against distance, Mississippi river at Greenville, Mississippi.

and amplitude to approximate the river data. A sine curve fitted to the crosses was used to generate a plot of channel direction against distance which has been superimposed on the map of the river as

River Meanders and the Theory of Minimum Variance 245

the dashed line in the upper diagram. There is reasonable agreement.

The same technique has been used in the other examples chosen to cover a range in shape from oxbows to flat sinuous form, and including a variety of sizes. Fig. 8.3B shows a stream that not only

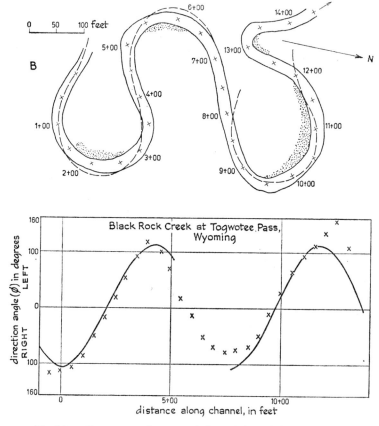

*Fig. 8.3*B *Sine-generated curve and plots of channel direction against distance, Black Rock Creek at Togwotee, Wyoming.*

is not as oxbowed as the Greenville bends, but also represents a river of much smaller size, the Mississippi river being about three-quarters of a mile wide and Black Rock Creek about 50 ft wide.

Fig. 8.3C shows the Paw Paw, West Virginia, bends of the Potomac river, similar to and in the same area as the great meanders of the tributary Shenandoah river. The curves in these rivers are characterised by being exceptionally elongated; amplitude is unusually large for the wave-length and both are large for the river width, characteristics believed to be influenced by elongation along the

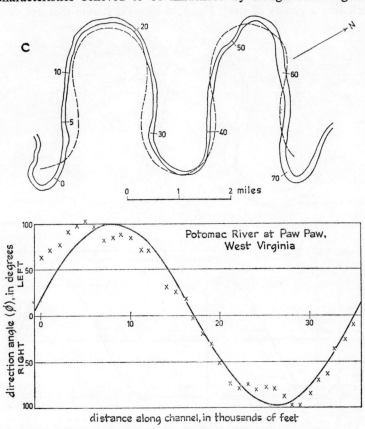

Fig. 8.3C Sine-generated curve and plots of channel direction against distance, Potomac river at Paw Paw, West Virginia.

direction of a fracture system in the rock (Hack and Young, 1959). Despite these peculiarities, the sine-generated curve fits the river well.

Figs. 8.3D and 8.3E (top) illustrate some of the best-known incised meanders in the western United States. These again provide plots of direction versus length which closely approximate to sine curves.

By far the most symmetrical and uniform meander reach we have

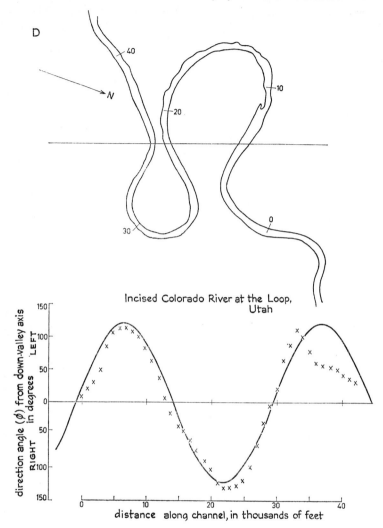

Fig. 8.3D Sine curve and plots of channel direction against distance, Colorado river at the Loop, Utah.

ever seen in the field is on the Popo Agie river near Hudson, Wyoming, for which flow data will be discussed later. The relation of channel direction to distance is compared with a sine curve in the lower part of Fig. 8.3E.

The laboratory experiments conducted by Friedkin (1945) provide an example of near-ideal meanders. The meander shown in Fig. 8.4 is one of those that developed greatest sinuosity. The channel shown

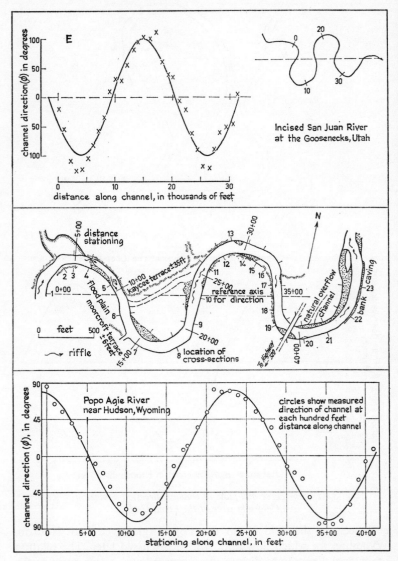

Fig. 8.3E Above: *Sine curve and plots of channel direction against distance, San Juan river at the Goosenecks, Utah.* Below: *Sine curve, Popo Agie river near Hudson, Wyoming.*

in Fig. 8.4A was developed in Mississippi river sand on a slope of 0·006, by rates of flow that varied from 0·05 to 0·24 cu. ft per sec. over an eighteen-hour period on a schedule that simulated the fluctuations of discharge on the Mississippi river. The sand bed was homogeneous and the meander mapped by Friedkin is more regular than those usually observed in natural streams.

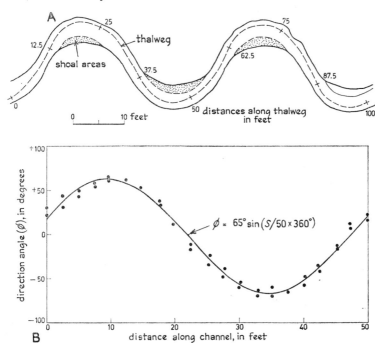

Fig. 8.4 Laboratory meander.
A, Plan. B, Comparison of direction angle with sine curve.

The direction angles of the thalweg were measured at intervals of 2·5 ft. These direction angles are plotted on the graph, Fig. 8.4B. Since each meander had a thalweg length of 50 ft, the data for the two meanders are shown on the same graph. Where the two bends had the same direction, only one point is plotted. The data correspond closely to the sine curve shown.

Comparison of variances of different meander curves
As a close approximation to the theoretical minimum, when a meander is such that the direction, φ, in a given unit length Δs

is a sine function of the distance along the curve, then the sum of squares of changes in direction from the mean direction is less than for any other common curve. In Fig. 8.5, four curves are presented.

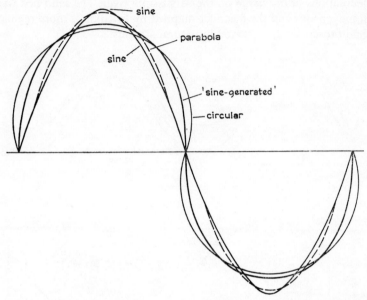

Fig. 8.5 Four symmetrical sinuous curves of identical wave-length and sinuosity.

One consists of joined portions of circles. Another is a pure sine curve; the third is a parabola. A fourth is sine-generated in that, as has been explained, the change of direction bears a sinusoidal relation to distance along the curve. All four have the same wavelength and sinuosity. However, the sums of squares of the changes in direction, as measured in degrees over ten equally spaced lengths along the curves, differ greatly, as follows:

Curve	$\Sigma(\Delta\varphi)^2$
Parabola	5,210
Sine curve	5,200
Circular curve	4,840
Sine-generated curve	3,940

The sine-generated curve has the least sum of squares. The theoretical minimum curve is identical within practical limits of drawing.

River Meanders and the Theory of Minimum Variance

Meander length, sinuosity and bend radius

The planimetric geometry of river meanders has been defined above by equation (8.10). The values of ω and M in this equation can be defined further, as follows:

The angle ω is a unique function of the sinuosity and is independent of the meander length. It ranges from zero for a straight-line path of zero sinuosity to a maximum of 125° for gooseneck meanders at point of incipient crossing. The sinuosity, k, equals the average of the values of $\cos \varphi$ over the range from $\varphi = 0$ to $\varphi = \omega$. Thus a relationship can be defined between k and ω. An approximate algebraic expression is

$$\omega(\text{radians}) = 2 \cdot 2 \sqrt{\frac{k-1}{k}} \qquad (8 \cdot 12\text{A})$$

or

$$\omega = 125° \sqrt{\frac{k-1}{k}} \qquad (8 \cdot 12\text{B})$$

Sinuosity, as measured by parameter k, is (as explained) thought to be a consequence of profile development which is only secondarily influenced by reaction from the meander development.

In the random-walk model, the standard deviation of changes in direction per unit of distance is σ, which has the dimension of L^{-1}, the reciprocal of length. As previously shown, σ is inversely proportional to meander length.

Bend radius is related to wave-length and is virtually independent of sinuosity. Defined as before, as the average over the $\frac{1}{6}$ of channel length for which φ is nearly linearly related to channel distance, bend radius R is

$$R = \frac{\frac{1}{6}M}{\Delta \varphi} \qquad (8 \cdot 13\text{A})$$

Since φ ranges from $+0 \cdot 5\omega$ to $-0 \cdot 5\omega$ over this near-linear range, $\Delta \varphi = \omega$. Substituting for ω its algebraic equivalent in terms of sinuosity,

$$\omega = 2 \cdot 2 \sqrt{\frac{k-1}{k}} \qquad (8 \cdot 13\text{B})$$

and since $M = k\lambda$, bend radius equals

$$R = \frac{\lambda}{13} \frac{k^{3/2}}{\sqrt{k-1}} \qquad (8 \cdot 13\text{C})$$

Some typical values for bend radius in terms of sinuosity are

Sinuosity	Bend radius (R)	Ratio wave-length / bend radius
1·25	0·215λ	4·6
1·5	0·20λ	5·0
2·0	0·22λ	4·5
2·5	0·24λ	4·2

These values agree very closely with those found by measurement of actual meanders. Leopold and Wolman (1960) describe meanders where sinuosities are dominantly between 1·1 and 2·0, with the average ratio of meander length to radius of curvature 4·7, which is equivalent to 0·213λ. The agreement with values listed above from the theory is satisfactory.

The curve defined by $\varphi = \omega \sin s/M\ 360°$ is believed to underlie the stable form of meanders. That actual meanders are often irregular is well known, but, as observed above, those meanders that are regular in geometry conform to this equation. Deviations (or 'noise') are it is surmised, due to two principal causes: (1) shifts from an unstable to a stable form caused by random actions and varying flow, and (2) non-homogeneities such as rock outcrops, differences in alluvium, or even trees.

Irregularities are more to be observed in free meanders than in incised meanders. During incision, irregularities apparently tend to be averaged out, and only the regular form is preserved.

MEANDERS COMPARED WITH STRAIGHT REACHES

Meanders are common, whereas straight reaches of any length are rare. It may be inferred, therefore, that straightness is a temporary state. As the usual and more stable form, according to the thesis of this report, the meander should be characterised by lower variances of the hydraulic factors, a property shown by Maddock (personal communication) to prevail in self-formed channels.

Review of the characteristics of meanders and the theories about them showed that quantitative data were nearly non-existent on the hydraulics of flow during those high stages or discharges at which channel adjustment takes place.

Because there are certain similarities and certain differences between straight segments or reaches of river and meandering reaches, it seemed logical to devote special attention to comparison of straight and curved segments during channel-altering flow. An initial attempt to obtain such data was made during 1954–8 on a small river in Maryland, but it became clear that rivers which rise and fall in stage quickly do not allow a two-man team with modest equipment to make the desired observations.

Attention was then devoted to the rivers which drain from the Wind River Range in central Wyoming, where good weather for field work generally exists during the week or two in late spring or early summer when snow-melt run-off reaches its peak. The plan of field work involved the selection of one or more meanders of relatively uniform shape which occurred in the same vicinity with a straight reach of river at least long enough to include two riffles separated by a pool section. No tributaries would enter between the curved and straight reaches, so that at any time discharge in the two would be identical. Bed and bank materials were apparently identical in the two reaches, although later bed sampling showed that there were some differences in bed-material size which had not been apparent to the eye.

A curve introduces an additional form of energy dissipation not present in a comparable straight reach, an energy loss due to change of flow direction (Leopold *et al.*, 1960). This additional loss is concentrated in the zone of greatest curvature midway between the riffle bars, at a location where, in a straight reach, energy dissipation is smaller than the average for the whole reach.

The introduction of river bends, then, tends to equalise the energy dissipation through each unit length, but does so at the expense of greater contrasts in bed elevation. Moreover curvature introduces a certain additional organisation into the distribution of channel-bed features. The field measurements were devised to measure and examine the variations in these quantities. Where the profiles of bed and water surface are plotted on the same graph for comparison (Fig. 8.6), it is apparent that at moderately high stage – from three-fourths bankfull to bankfull – the larger-scale undulation of the river bed caused by pool and shallow has been drowned out, no longer causing an undulation in the water-surface profile of the meander. In contrast, at the same stage in the straight reach the flat and steep alternations caused by pool and riffle are still discernible. On the other hand the profile reveals that the undulation of the bed along the stream length is of larger magnitude in the curved than in the straight reach.

254 *Walter B. Langbein and Luna B. Leopold*

One could reason as follows: in a straight reach of channel the dunes, bars, pools and riffles form more or less independently of the

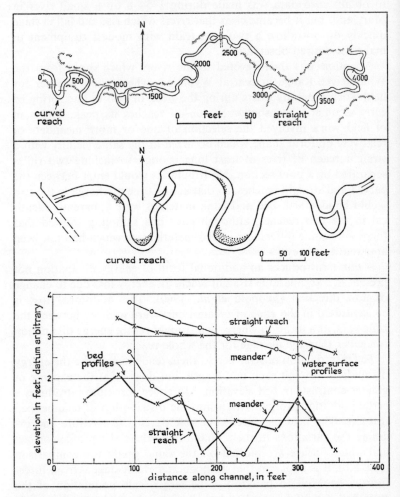

Fig. 8.6 Plan and profiles of straight and curved reaches on Baldwin Creek, near Lander, Wyoming: in lower diagram, curved reach shown by pecked lines and straight reach by solid lines.

channel pattern owing to grain interaction; to the occurrence of pool and riffle the channel must adjust; the riffle causes a zone of greater-than-average steepness which is also a zone of greater-than-average energy expenditure; the pool on the other hand, being

River Meanders and the Theory of Minimum Variance

relatively free of large gravel on the bed and of larger depth relative to bed roughness elements, offers less resistance to flow and there the energy expenditure is less than average. The result is a stepped water-surface profile – flat over pool and steep over riffle.

Once pools and riffles form with their consequent variations in depth, width remaining relatively uniform, then slopes must vary. If slopes vary so that bed shear remains constant, then slope varies inversely with depth. However, this result would require the friction factor to vary directly as the square of the depth. If, on the other hand, the friction factor remained uniform and the slope varied accordingly, then bed shear would vary inversely as the square of the depth. Uniformity in bed shear and the friction factor is incompatible, and the slope adjusts so as to accommodate both about equally and minimise their total variance.

Field measurements
When a satisfactory site was located, a continuous water-stage recorder was established; bench marks and staff gauge were installed and the curved and straight reaches mapped by plane table. Cross-sections were staked at such a spacing that about seven would be included in a length equal to one pool-and-riffle sequence.

Water-surface profiles were run by levelling at one or more stages of flow. Distances between water-surface shots were equal for a given stream; 10-ft distances were used on small streams having widths of 10 to 20 ft, and 25-ft spacing used for streams 50 to 100 ft wide. Usually two rods, one on each stream bank, accompanied the instrument. Shots were taken opposite one another along the two banks. Water-surface elevations were read to 0·01 ft on larger rivers, and to 0·001 ft on small ones. For the small rivers an attachment was used on a surveying rod which in construction resembled a standard point gauge used in laboratory hydraulic practice for measurement of water-surface elevation.

Velocity measurements by current meter were usually made at various points across the stream at each cross-section or at alternate ones.

Figs 8.6, 8.7 and 8.8 show examples of the planimetric maps and bed and water-surface profiles. The sinusoidal change of channel direction along the stream length for Fig. 8.7 has been presented in Fig. 8.3.

Reduction of data

Mean depths and mean velocities were calculated for each cross-section. The average slopes of the water surface between the cross-sections were also computed from the longitudinal profile. To reduce

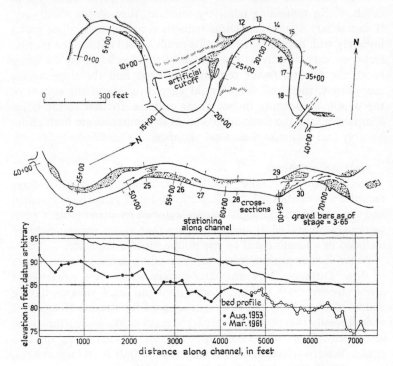

Fig. 8.7 Plan and profile of curved and straight reaches on the Popo Agie, near Hudson, Wyoming.

these quantities to non-dimensional form, they were expressed in ratio to the respective means over each reach. The variances of these ratios were computed by the usual formula

$$\sigma_x^2 = \frac{\Sigma X^2}{N} - \left(\frac{\Sigma X}{N}\right)^2 \tag{8.14}$$

where σ_x^2 is the variance of the quantity X and N is number of measurements in each reach.

As has been shown by Maddock, the behaviour of many movable-bed streams can be explained by a tendency towards least variation in bed shear and in the friction factor. Accordingly, the variances of the bed shear and of the friction factor – again considered as ratios to their respective means – were also computed. Bed shear

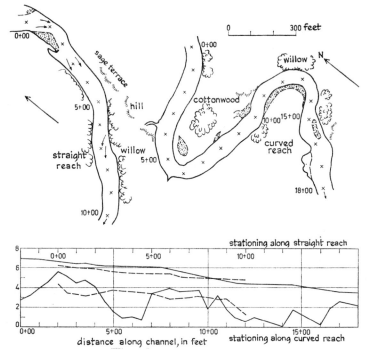

Fig. 8.8 Plan and profiles of straight and curved reaches on Pole Creek, near Pinedale, Wyoming: in lower diagram, curved reach shown by pecked lines and straight reach by solid lines.

is equal to the product γDS, where γ is the unit weight of water, D is the depth and S is the slope of the energy profile. The unit weight of water is a constant that may be neglected in this analysis of variances. In this study water-surface slopes will be used in lieu of slopes of the energy profile. This circumstance involves an assumption that the velocity head is small relative to the accuracy of measurement of water surface. In any case, where the velocity-head corrections were applied, they were either small or illogical, and so were neglected.

Similarly, the variance of the quantity DS/v^2, which is proportional to the Darcy–Weisbach friction factor, was computed.

Width is not included in the analysis because it is relatively uniform, and lacks distinctive characteristics that vary from curved to straight reaches.

Comparison of results

The results are listed in Table 8.1. The variance of slope in each reach is larger than that of depth or velocity. The contrast, however, is less in the meandering than in the straight reach; this reduction may be noted by the figures in the column headed *a*, which is the ratio of the variance of slope to the sum of those of depth and velocity. Among the five rivers studied, this ratio is uniformly less in the meandering reaches than in the straight reaches.

TABLE 8.1

Sinuosity and variances for straight and curved reaches on five streams

Stream	Sinuosity (k)	σ_D^2	σ_v^2	σ_S^2	a	$\sigma_{DS}^{'2}$	σ_f^2
Straight reaches							
Pole Ck nr Pinedale, Wyo.	1·05	0·019	0·042	0·26	4·2	0·25	0·21
Wind R. nr Dubois, Wyo.	1·12	0·03	0·017	0·16	3·4	0·14	0·11
Baldwin Ck nr Lander, Wyo.	1·00	0·065	0·058	0·94	7·8	0·55	0·40
Popo Agie R. nr Hudson, Wyo.	1·02	0·026	0·017	0·70	16·0	0·46	0·30
Mill Ck nr Ethete, Wyo.	1·02	0·053	—	0·28	2·6	0·32	—
Curved reaches							
Pole Ck nr Pinedale, Wyo.	2·1	0·067	0·088	0·41	2·6	0·15	0·07
Wind R. nr Dubois, Wyo.	2·9	0·074	0·032	0·13	1·2	0·06	0·073
Baldwin Ck nr Lander, Wyo.	2·0	0·048	0·017	0·20	3·1	0·15	0·14
Popo Agie R. nr Hudson, Wyo.	1·65	0·044	0·029	0·38	5·2	0·28	0·12
Mill Ck nr Ethete, Wyo.	1·65	0·145	—	0·64	2,2	0·21	—

Sinuosity: ratio of path distance to down-valley distance. Variances: σ_D^2, depth; σ_v^2, velocity; σ_S^2, slope; σ_{DS}^2, bed shear; σ_f^2, Darcy–Weisbach friction factor.

The data listed in Table 8.1 also show that the variance in bed shear and in the friction factor are uniformly lower in the curved than in the straight reach. The data suggest that the decrease in the variance of bed shear and the friction factor is related to the sinuosity. As shown in Fig. 8.9, the greater the sinuosity the less the average of these two variances.

When one considers, for example, that the variances of depth and velocity are greater in the curved reach, and even that the variance of slope may be greater (as for example Pole Creek), the fact that the

River Meanders and the Theory of Minimum Variance 259

product DS and the ratio DS/V^2 have consistently lower variances must reflect a higher correlation between these variables in the

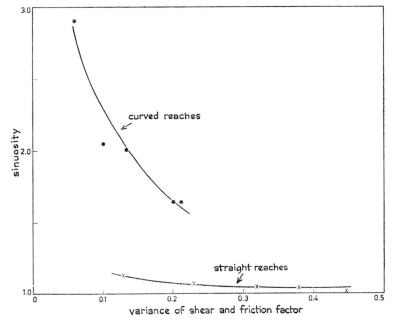

Fig. 8.9 Relation of sinuosity to average of variances of friction factor and shear.

meandered than in the straight reach. This improved correlation is shown in the tabulation (Table 8.2) of the respective coefficients of determination (=square of the coefficient of correlation).

TABLE 8.2

Comparison of coefficients of determination

	Straight reach		Curved reach	
	r_{DS}^2	r_{vS}^2	r_{DS}^2	r_{vS}^2
Pole Creek	0·044	0·270	0·72	0·90
Wind River	0·072	0·062	0·55	0·42
Baldwin Creek	0·56	0·28	0·42	0·72
Pope Agie River	0·39	0·49	0·58	0·62
Mill Creek	0·00	–	0·88	–

r_{DS}^2, coefficient of linear determination between depth and slope;
r_{vS}^2, coefficient of linear determination between velocity and slope.

Fig. 8.10 shows graphical plots for the Pole Creek data. The graph shows that, in the meanders, depths, velocities and slopes are more systematically organised than in straight reaches.

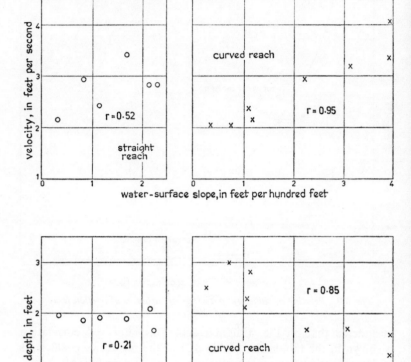

Fig. 8.10 *Velocity, depth and slope on Pole Creek near Pinedale, Wyoming.*

To summarise, the detailed comparisons of meandering and straight reaches confirm the hypothesis that in a meander curve the hydraulic parameters are so adjusted that greater uniformity (less variability) is established among them. If a river channel is considered in a steady state, then the form achieved should be such as to avoid concentrating variability in one aspect at the expense of another. In particular a steady-state form minimises the variations in forces on the boundary – that is, the steady-state form tends towards more uniform distribution of both bed stress and the friction factor.

INTERPRETATIVE DISCUSSION

After a review of data and theories concerning river meanders, Leopold and Wolman (1960) concluded that meander geometry is related in some unknown manner to a more general mechanical principle (p. 774). The status of knowledge suggested that the basic reason for meandering is related to energy expenditure, and they concluded (p. 788) as follows: 'Perhaps abrupt discontinuities in the rate of energy expenditure in a reach of channel are less compatible with conditions of equilibrium than is a more or less continuous or uniform rate of energy loss.'

Subsequent work resulted in the postulate that the behaviour of rivers is such as to minimise the variations in their several properties (Leopold and Langbein, 1962); the present work shows how this postulate applies to river meanders as had been prognosticated.

Total variability cannot be zero. A reduction in the variability in one factor is usually accompanied by an increase in that of another. For example, in the meander, the sine-generated curve has greater variability in changes in direction than the circle. This greater variability in changes in direction is, however, such as to decrease the total angular change. The sine-generated curve, as an approximation to the theoretical curve, minimises the total effect.

The meandering river has greater changes in bed contours than a straight reach of a river. However, these changes in depth produce changes in the slope of the water surface and so reduce the variability in bed shear and in the friction factor, as well as lessening the contrast between the variances of depth and slope.

These considerations lead to the inference that the meandering pattern is more stable than a straight reach in streams otherwise comparable. The meanders themselves shift continuously; the meandering behaviour is stable through time.

This discussion concerns the ideal case of uniform lithology. Nature is never so uniform, and there are changes in rock hardness, structural controls and other heterogeneities related to earth history. Yet despite these natural inhomogeneities, the theoretical forms show clearly.

This does not imply that a change from meandering to straight course does not occur in nature. The inference is that such reversals are likely to be less common than the maintenance of a meandering pattern. The adjustment towards this stable pattern is, as in other geomorphic processes, made by the mechanical effects of erosion and deposition. The theory of minimum variance adjustment describes the net river behaviour, not the processes.

REFERENCES

FRIEDKIN, J. F. (1945) *A Laboratory Study of the Meandering of Alluvial Rivers* (Vicksburg, Miss., U.S. Waterways Expt. Sta.) 40 pp.

HACK, J. T., and YOUNG, R. S. (1959) *Intrenched Meanders of the North Fork of the Shenandoah River, Va.*, U.S. Geol. Survey, Prof. Paper 354–A, 1–10.

LANGBEIN, W. B. (1964) 'Geometry of river channels', *J. Hydrol. Div., Amer. Soc. Civ. Eng.*, 301–12.

LEOPOLD, L. B., BAGNOLD, R. A., WOLMAN, M. G., and BRUSH, L. M. (1960) *Flow Resistance in Sinuous or Irregular Channels*, U.S. Geol. Survey, Prof. Paper 282–D, 111–34.

——, and LANGBEIN, W. B. (1962) *The Concept of Entropy in Landscape Evolution*, U.S. Geol. Survey, Prof. Paper 500–A, 20 pp.

——, and WOLMAN, M. G. (1957) *River Channel Patterns: Braided, Meandering, and Straight*, U.S. Geol. Survey, Prof. Paper 282–B, 84 pp.

—— (1960) 'River meanders', *Bull. Geol. Soc. Amer.*, LXXI 769–94.

VON SCHELLING, H. (1951) 'Most frequent particle paths in a plane', *Trans. Amer. Geophys. Union*, XXXII 222–6.

—— (1964) *Most Frequent Random Walks* (Schenectady, N.Y., Gen. Elec. Co. Rept. 64GL92).

APPENDIX: SYMBOLS

- a ratio of variances (Table 8.1)
- c a coefficient (equation (8.1))
- k sinuosity, ratio of path distance to down-valley distance
- R ratio of curvature of a bend
- r coefficient of correlation (Table 8.2)
- s unit distance along path
- x a factor or variable
- p probability of a particular direction (equation (8.1))
- σ^2 variance of a particular factor, x
- D water depth
- S energy slope
- M total path distance in a single wave-length
- N number of measurements
- v velocity
- $\propto$ a constant of integration (equation (8.2))

γ   unit weight of water
λ   wave-length
π   pi, ratio of circumference to radius
ρ   radius of curvature of path
σ   standard deviation
φ   angle that path at a given point makes with mean down-path direction
ω   maximum angle a path makes with mean down-path direction

9 General Theory of Meandering Valleys and Underfit Streams

G. H. DURY

THE bends of winding valleys can range from highly irregular to distinctly systematic. Little if any attention appears to have been directed to irregularly winding valleys, whereas systematically winding valleys, under the title *meandering valleys*, have attracted considerable notice. Numbers of such valleys contain flood plains, on which present-day rivers describe free meanders far less ample than the enclosing valley bends (Fig. 9.1). Streams in such situations

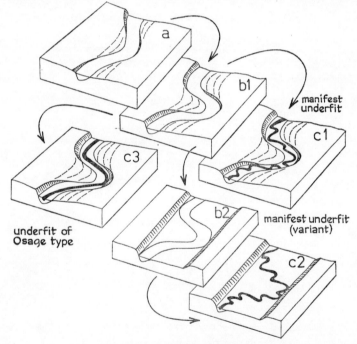

Fig. 9.1 Partial model for the development of underfit streams.

are commonly called *underfit*, the implication being that the windings of the valleys are authentic meanders – frequently, ingrown

General Theory of Meandering Valleys and Underfit Streams

meanders – of former large streams, that there is a relationship between stream size and size of meanders, and that some cause has operated to reduce stream size and meander size significantly below their former values. As will be shown later, it is convenient to distinguish meandering streams contained in more amply meandering valleys as *manifestly underfit*, since underfit streams of other kinds are also known.

Although some early discussions of meandering valleys and manifestly underfit streams sought to demonstrate the effects on stream pattern of bedrock structure or of erosion at times of overbank discharge, a widely held view has been that of Davis (1896, 1899, 1906, 1913), namely, that underfitness results from capture, the beheaded stream undergoing shrinkage and reducing the size of the meanders which it describes. Davis referred stream shrinkage in the general case to the captures expectable during the competitive development of rivers, choosing his leading examples from scarpland country where subsequents extend themselves along the strike of weak outcrops. It is true that he produced for the English Cotswolds a possible alternative hypothesis of the former discharge of spill water from an ice-dammed lake, and that he made a passing mention of the superposition of the effects of climatic change upon those of capture. However, overspill cannot be made to produce the observed effects in the Cotswolds; the suggestion of climatic change was not pursued in Davis's later writing, where (1923) the capture hypothesis was allowed to stand.

Let it be assumed that manifest underfitness results from reduction of stream volume; some regionally applicable hypothesis is then required. For despite Davis's denial, underfits are developed on the regional scale, and are so developed in those regions from which he took specific examples. Now it is impossible for all streams in a region to have been reduced by beheading, but for none of them to reveal the results of its own successful piracy. Even in well-documented cases of capture, diversion can be proved incapable of explaining more than a fraction of the total shrinkage. Similarly, the onset of manifest underfitness has been disentangled in a number of instances from the cessation of discharge of overspill or melt water. The required explanation of underfitness in the general case can hardly be other than climatic.[1]

[1] References to discussions of this and related matters are given in U.S. Geological Survey Professional Paper 452, where the topics are additionally explored. In order to avoid inflating the reference list at the end of the present chapter, citations of the writer's work are almost everywhere omitted.

Furthermore, this explanation must apply very broadly indeed. Underfit streams have been identified widely in Europe and as far east as the Ukraine. They occur in all the major climatic regions of the conterminous U.S.A. Little is known of them as yet in the southern hemisphere, although they are numerous in the rather special setting of the Riverine Plain of Australia (Langford-Smith, 1960); they are present also in the coastal drainage of Australia's Northern Territory (Hays, 1967) and on the eastern coastland. Underfits in the mid-latitude belts should be regarded not as exceptional but as normal. Although little is known of their distribution in very high or very low latitudes, they have at least been identified as far as 65°N. in Alaska, as far as 18°N. in Puerto Rico, and between 25° and 16°S. in Australia.

FIELD INVESTIGATIONS

In order to test the assumption that valley bends are authentic ingrown meanders of former large streams, the writer in 1951 began the sub-surface exploration of the bottoms of valleys occupied by manifest underfits. It seemed possible that the channels of former large streams, if such had existed, might in places be preserved. The first stream selected was the Warwickshire Itchen, which is well suited as a test of hypotheses additional to that of capture. If anything, the scarp-draining Itchen is extending its catchment at the expense of competitors on the backslope. The minutely permeable alluvium and wholly impermeable bedrock of the valley bottom prevent loss to percolation, whether deep or shallow; they also make irrelevant questions of increases in surface run-off on account of frozen ground. Discharge of melt water or spill water cannot be held to explain the valley meanders of the Itchen. Not only does the stream flow towards the direction of former local ice fronts; the valley meanders were still being shaped after the last local ice had disappeared, and after a whole interglacial and the succeeding glacial had supervened.

Drilling into the flood plain established the presence beneath it of a large channel which meanders round the valley bends, with asymmetrical pools at the extremities of these bends and shallower and more regular cross-sections between bends. The base of the channel in its mid-section lies deeper than the depth of maximum scour in the pools of stream meanders. Similar exploration of additional valleys produced generally similar results, although at two sites the reactivated present stream has reached bedrock and is eroding it. The Cherwell, a direct competitor of the Itchen, is also underfit; it too possesses a large filled channel. Among the Cotswold streams on which large channels have been proved, the Coln has special significance. It was used by Davis not only as a leading example of underfitness resulting from capture, but as an instance of the successive capture of two headstreams whereby Davis supposed the Coln to have undergone two successive reductions. However, the features identified by Davis as indicating the second capture do not exist on the ground; and the Coln is no more, and no less, underfit than its neighbours and competitors.

The earliest series of field observations provided an average of 11·5/1 for the ratio W/w between the respective bed widths of former and present-day channels, and an average of 13/1 for the ratio L/W between the wave-length of valley meanders and the bed width of large channels. Observed values of l/w, the wave-length/width ratio on existing meandering streams, produce a mode of about 10/1 which accords with the predictions of the theory of flow around bends (Leopold and Wolman, 1960; Bagnold, 1960). However, the indicated average of 13/1 for L/W is within the range of values obtained for the corresponding ratio on existing streams. The ratio W/w provides an index of underfitness. But since values of L are more easily to be had than values of W, the alternative index given by the wave-length ratio L/l may be substituted. A group of readings in Lowland England places L/l in the range between 9/1 and 10/1. Hydrological implications of these findings are discussed later.

Fieldwork in the U.S.A. during 1960–1, in addition to supplying numerous and widely distributed examples of manifest underfits, showed that a value for L/l of about 5/1 is fairly common. Near the former ice fronts in Wisconsin the ratio ranges as high as 10/1, whereas in the Ozarks it can be as low as 3/1, having fallen off in a given catchment in the downstream direction.

Whereas manifest underfits are typical of Lowland England, several upland areas of the U.S.A. – for instance, the Driftless Area, the Ozarks and the Appalachians – contain ingrown meandering valleys where the curves of existing channels approximately reproduce the pattern of valley bends. In such cases the apparent value of l/w is high, e.g. between 40/1 and 50/1; a specific instance is the North Fork of the Shenandoah, in the reach frequently used to illustrate meanders of very high amplitude, where l/w is apparently 47/1. The bends in question are, however, those of the valley, not of the stream. In its upstream sections, and also on numbers of its laterals, the North Fork of the Shenandoah exhibits the usual value of about 10/1 for the ratio between wave-length and bed width. Some New England streams, curving round the large bends of deeply cut narrow-bottomed valleys, produce meanders appropriate in size to their present channels as soon as they emerge on to expanses of alluvium or outwash on wider valley floors. Some members of the Humboldt system in the Great Basin have stream meanders where they traverse basins, but pierce the fault blocks of the ranges in winding canyons where the curves are much greater in dimension.

The Osage river in the Ozarks serves to explain the apparent anomaly of inordinately great wave-length/width ratios in upland country. The survey of its bed profile during site exploration for Bagnell Dam, which now impounds the Lake of the Ozarks, reveals a marked sequence of pool and riffle. The average spacing of 0·56 miles from pool to pool would have matched a wave-length of 1·12 miles, had the stream been a meandering one. The average wave-length of valley meanders is distinctly greater, at 3·8 miles; and the ratio between this wave-length and the average double pool spacing is about 3·5, within the range obtained for the ratio L/l of other streams in the same area. Upstream of the Lake of the Ozarks the water surface of the Osage at extreme low stages lies well below the crests of riffles, revealing not one but three or four of these spaced along a single limb of the curved trace. Spacing is appropriate to the size of the existing channel, at about five bed widths between pool and pool or riffle and riffle. Except for being forcibly curved around the valley bends, the channel of the Osage behaves as if it were straight (Fig. 9.2). Reduction of channel size from the dimensions associated with the development of valley meanders has been followed by the development of pool and riffle on the new channel, but not by the appearance of stream meanders. The apparently high ratio l/w is in actuality the ratio L/w between wave-length of valley meanders and bed width of existing streams.

The channel pattern displayed on the Osage differs strongly enough from that of manifest underfits, and appears sufficiently widespread, to deserve separate sub-classification, and the writer has proposed the style *underfits of the Osage type* in a study of the Colo river of New South Wales (Dury, 1966), which in part belongs to this sub-class.

A third variation on the theme of underfitness is provided by the upper Evenlode. This is manifestly underfit, to the extent that it preserves the trace of former large meanders in addition to the existing stream meanders. However, by contrast with its middle reaches which flow through an ingrown meandering valley, the upper Evenlode is enclosed by sub-parallel scarps of caprock, below which

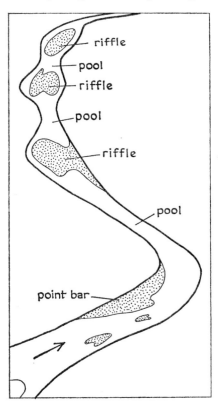

Fig. 9.2 Pool-and-riffle sequence on valley bends of the Osage river: perspective sketch from an air photograph.

weaker formations have been eroded into a broad trough. This is the meander trough of the former large stream. It seems uncommon for the two sets of meanders to be present on the floor of an open valley. Factors in the preservation of the earlier and larger set on the upper Evenlode may be the self-cemented limestone sludge-gravel into which the former large meandering channel was cut, and the lack of alluviation which could have raised the base of the existing channel above the level of the former resistant banks. There is room in the channel fill for the meanders and flood plain of the present stream, but the cemented gravel may well act in the same constraining manner as the bedrock walls of an ingrown valley.

Manifestly underfit streams in ingrown meandering valleys, their variant exemplified by the upper Evenlode, and underfits of the Osage type incompletely represent the possible range of underfit forms. At least eight types of end-product are imaginable, produced by one or other of eleven different modes of development; but the identification of the five end-types not discussed here is a matter for detailed sub-surface exploration not as yet undertaken in relevant settings.

CHRONOLOGY

The temporal relationships of underfit streams have two main aspects: the date of origin of the rivers by which valley meanders have been cut, and the date of the shrinkage by which underfit rivers were reduced to the underfit state. The latest date of origin for ingrown valley meanders is, in many areas, the time when incision of plateaux began. This may well also be the earliest date of origin, for there is no reason to look on ingrown valley meanders as invariably inherited from free meanders on a flood plain at high level; the record of ingrowth is necessarily one of increasing sinuosity, and the depth/sinuosity characteristics of numerous valleys suggest that sinuosity was very low indeed when incision started.

Not much is known for certain of the stratigraphical position of planated plateau surfaces; but Troll (1954) has shown that the incised meandering valleys of the Alpine Foreland and the Hercynian massifs of Europe were initiated early in the Pleistocene, while Kremer (1954) refers the highest terrace of the ingrown Moselle to infilling during the Gunz (=Early, Nebraskan) Glacial. There is no doubt that ingrowing valley meanders existed on the Moselle as early as the Mindel (=Antepenultimate, Kansan) Glacial. Later examples of initiation of meandering valleys include the Warwickshire Avon, which came into being shortly after the recession of the Penultimate (=Riss, Illinoian) ice. A general sequence may be provided by the South Atlantic Coastal Plain of the U.S.A., where Doering (1960) identifies six off-lapping formations as ranging in age from early pre-glacial Pleistocene to mid-Wisconsin Glacial; all are trenched by ingrown meandering valleys, suggesting that the total history of the initiation and development of these valleys spans

General Theory of Meandering Valleys and Underfit Streams 271

the whole of Pleistocene time. It need not be inferred, however, that the high former discharges responsible for shaping valley meanders operated throughout the whole of the interval concerned; on the contrary, repeated episodes of shrinkage to underfitness are envisaged in the theory now being developed.

Dates obtained for the last major shrinkage of streams which are now underfit fall within a narrow range of absolute time. The fossil pollen in the channel fill of the Cotswold Dorn shows that filling had begun by 9000 B.P. Valley meanders on the Shropshire–Cheshire Plain must postdate the till sheet into which they are cut; an approximate date for the local disappearance of ice is given by the age of 11,000 years determined for peat in a kettle-hole. Valley meanders on the floor of Glacier Lake Hitchcock in the Connecticut valley are similarly dated to about 12,000 B.P. Two fixes are given by Black Earth Creek, Wisconsin. This manifestly underfit stream occupies, as a shallow meander trough cut into loess, the slightly ingrown but highly sinuous channel of a former and larger stream. Its valley is crossed by the outermost moraine of the Wisconsin (=Last, Wurm) ice sheet; the ice stand belongs in the interval 14,000–12,000 B.P. Ice recession was followed by loess fall. Only after the loess sheet had been spread did valley meanders come into being. The fill of the large channel, below the existing flood plain, is barren of pollen, but the base of this flood plain has been dated by the reference of its pollen content to a controlling radiocarbon scale to about 10,000 B.P., by which time Black Earth Creek had already been reduced to underfitness.

Evidence from river terraces, less satisfactory in some ways than that from channel fills, also places the last main shrinkage well within Last-deglacial time. Somewhat scanty additional evidence from a few valleys, plus the comparison of the fluvial record with the record of pluvial lakes in the Great Basin and elsewhere, suggests that this shrinkage may have been double. In the interval corresponding to pollen zone Ic (Older Dryas), i.e. between about 12,750 and 12,000 B.P., the pluvial lakes stood very high and certain valleys appear to have been scoured to the maximum. Between about 12,000 and 10,000 B.P. pluviosity markedly decreased; in this interval belong P.Z. II (Allerød) and the Two Creeks Interstade of North America. Subsequent increased pluviosity restored high lakelevels between about 10,000 and 9000 B.P., during P.Z. IV (pre-Boreal), which has long been known by its paleontological and floristic record as a time of increasing dryness. This appears to have been the time when general underfitness was finally confirmed, except in areas which were still covered by ice or by pro-glacial lakes.

The net record since 9000 B.P. is one of desiccation, although the change has proved uneven and at times partly reversible. Partial re-excavation of some channel fills has been proved for P.Z. VII (main Atlantic), between about 6200 and 4500 B.P., and a few sites provide complex histories of episodic channelling at other times since 9000 B.P., which, however, also failed to remove the whole of the fill.

Areas which were covered by ice or pro-glacial lakes as late as 9000 B.P. can nevertheless contain underfit streams; examples are the catchments of the Fox and East River systems draining into the Green Bay arm of Lake Michigan, which still lay beneath the waters of an enlarged Lake Erie as late as 2500 or even 2000 B.P. It is not yet certain how the valley meanders here fit into the sequence identified for areas more remote from the centres of ice dispersion. They may correspond to one of the lesser and later episodes of general channelling elsewhere; alternatively, they may record the special meteorological and climatic conditions of districts not very far from a receding ice front.

IMPLICATIONS

Argument so far has been made to rely on the general assumption that there is a relationship between stream size and size of meanders. Expressed in these terms, the assumption is not meaningful. For immediate purposes it may be permissible to forgo a discussion of the concept of channel-forming discharge, and merely to state that there is copious evidence for the proposition that, as a generality, meander wave-length and stream bed width vary as the square root of discharge at bankfull (Fig. 9.3). There is also reason to suppose that channel-forming discharge = bankfull discharge = discharge at the recurrence interval of 1·58 years on the annual flood series, i.e. discharge at the most probable annual flood (Dury *et al.*, 1963; Dury, 1968). Accordingly, stream shrinkage should be defined as significant reduction in discharge at the recurrence interval corresponding to channel-forming flow; or, more economically, as significant reduction in discharge at the most probable annual flood.

The general relationship $l, w \propto q^{0.5}$ leads as a first approximation to the proposition $Q/q = (L/l)^2$, where Q, q are respectively the channel-forming discharges of the former large streams and of the

General Theory of Meandering Valleys and Underfit Streams

corresponding present-day underfits. This would give Q/q a value of about 100/1 in Lowland England and 25/1 in large areas of the U.S.A. Allowance for contrasts between former and present channels in respect of cross-sectional form, downstream slope,

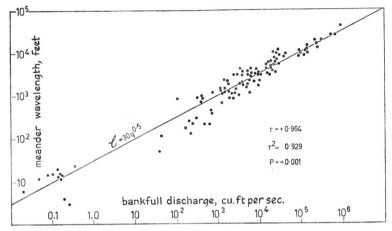

Fig. 9.3 Relationship of wave-length to bankfull discharge: best-fit equation obtained by computer analysis.

roughness and velocity suggest that the discharge ratio is somewhat less than that computed merely from the wave-length ratio, being perhaps 50/1 or 60/1 for high values of L/l and about 20/1 or less where L/l is 5/1.

The writer's first attempts (Dury, 1954) to derive the order of precipitation which could have nourished streams capable of shaping valley meanders are vitiated by their dependence on the rational formula of run-off. More refined calculations, using transformations of the precipitation–temperature–run-off relationships defined by Langbein et al. (1949), suggest that the channel-forming discharges required to explain the former channel patterns of streams which are now underfit could have been provided, in the ranges of time indicated earlier, by increases in mean annual precipitation of 50 to 100 per cent. The demand for increased precipitation, which added its influence to the increase in surface moisture resulting from reduced evapotranspiration, matches the demand made in reconstructions of the history of pluvial lakes in non-glaciated regions. Magnitude–area–intensity analysis of precipitation shows that the most economical means of securing the required increase in precipitation is an increase in the frequency and power of frontal storms. In this way the general theory of meandering valleys comes to imply an increase in storminess, not only in mid-latitudes but also beyond, during the last episode of deglaciation. The state of the global circulation during maximum glaciation remains still uncertain. Pluvial lakes supply evidence of high pluviosity during the Last-glacial maximum, but lake stands then appear to have been lower than they were in the deglacial intervals signalised for the last scouring of large channels in meandering valleys. On the general ground that one deglacial episode is likely to have followed the pattern of earlier deglacials, it can be provisionally assumed that the stream shrinkages in the range 12,000 to 9000 B.P. were no more than repetitions of similar events in corresponding earlier brackets of the Pleistocene sequence.

REFERENCES

BAGNOLD, R. A. (1960) *Some Aspects of the Shape of River Meanders*, U.S. Geol. Survey, Prof. Paper 282–E, 10 pp.
DAVIS, W. M. (1896) 'La Seine, la Meuse, et la Moselle', *Ann. de Géogr.*, VI 25–49.
—— (1899) 'The drainage of cuestas', *Proc. Geol. Assoc. Lond.*, XVI 75–93.
—— (1906) 'Incised meandering valleys', *Bull. Geol. Soc. Philadelphia*, IV 182–92.
—— (1913) 'Meandering valleys and underfit rivers', *Ann. Assoc. Amer. Geogr.*, III 3–28.
—— (1923) 'The cycle of erosion and the summit level of the Alps', *J. Geol.*, XXXI 1–41.
DOERING, J. A. (1960) 'Quaternary surface formations of the southern part of the Atlantic Coastal Plain', *J. Geol.*, LXVIII 182–202.
DURY, G. H. (1954) 'Contribution to a general theory of meandering valleys', *Amer. J. Sci.*, CCLII 193–224.
—— (1964a) *Principles of Underfit Streams*, U.S. Geol. Survey, Prof. Paper 452–A, 67 pp.
—— (1964b) *Subsurface Explorations and Chronology of Underfit Streams*, U.S. Geol. Survey, Prof. Paper 452–B, 56 pp.
—— (1965) *Theoretical Implications of Underfit Streams*, U.S. Geol. Survey, Prof. Paper 452–C, 43 pp.
—— (1966) 'Incised valley meanders on the lower Colo River, New South Wales', *Austr. Geographer*, X 17–25.
—— (1968) 'Bankfull discharge and the magnitude-frequency series', *Austr. J. Sci.*, XXX 371.
——, HAILS, J. R., and ROBBIE, M. B. (1963) 'Bankfull discharge and the magnitude-frequency series', *Austr. J. Sci.*, XXVI 123–4.
HAYS, J. (1967) 'Land surfaces and laterites in the north of the Northern Territory', in J. N. Jennings and J. A. Mabbutt (eds), *Landform Studies from Australia and New Guinea* (Canberra, A.N.U. Press) 182–210.
KREMER, E. (1954) 'Die Terrassenlandschaft der mittleren Mosel . . .', *Arbeit. z. Rhein. Landeskunde*, VI.

LANGBEIN, W. B., et al. (1949) *Annual Runoff in the United States*, U.S. Geol. Survey, Circ. 52, 14 pp.
LANGFORD-SMITH, T. (1960) 'The dead river systems of the Murrumbidgee', *Geogr. Rev.*, L 368–89.
LEOPOLD, W. B., and WOLMAN, M. G. (1960) 'River meanders', *Bull. Geol. Soc. Amer.*, LXXI 769–94.
TROLL, C. (1954) 'Über Alter und Bildung von Talmaändern', *Erdkunde*, VIII, no. 4, 286–302.

Index

ability to corrade, 97
abstraction, 114
acceleration, 110
accordance of elevation of point bar, 177
accretion, 24–5; lateral, 171–2, 178; vertical, 171 ff.
accretion deposits, 25
accumulation of disintegrated rock, 98
adjustments in natural channels, 229–231
aggradation, 197–8, 202, 205, 207, 222, 228, 231
amplitude, 55, 246
anastomosis, 197–8
angle of attack, 23
annual flood, 166, 177
antecedence, 115
antecedent: slope, 158; surface, 159
area: of catchment, 78; of contact, 103
arid regions, 99–100
assumed order, 130–2

bank: erosion of, 44; material, 23; rockbound, 31
bankfull discharge, bankfull flow, 171, 214, 272
bankfull stage, 166, 231, 253
bar, 166, 198–201, 203, 206, 254; central, 200–1, 203, 205–6; coalescent, 177; deposition, hypotheses of, 202; incipient, 205; linear, 201, 205; rock, 29; stabilised, 201–2; submerged, 203
base-level, 59, 74 ff.; ancient, 84, 93; ancient, values for, 85–7; change in, 87; estimate of, 90; temporary, 76
base-levelling, 77
basin characteristics, 184
bed: condition, 233; configuration, 225, 231; contours, 261; elevation, 253; load, 151, 200; material size, 223; profile, 268; shear, 255–7; transport, 203

beheading, 265
belt, of no erosion, 145–9, 156–62; marginal, 156
bend radius, 251–3
bifurcation ratio, 125, 129–33
birth of theory, 68
blows, 105
bluffs, 44
braid, braiding, 217–19, 235; development of, 200–11
braided river, 195, 211
branching streams, 107–8
breaking down of divides, 156
Brownian movement, 151
bulk of terrace deposits, 40

calibre, 24–5, 78, 88, 97, 100–2, 104, 173–5, 183, 200, 219, 234. See also grain size
capacity, 60–1, 64, 102, 104–7, 234; inequality of, 106
capture, 265, 267
carbon-14 dates, 185, 189, 193, 271
cataclysmal views, 19
caving of banks, 205
change: in base-level, 87; in gradient, 76; of sea-level, 41; of temperature, 95–7; of volume, 64
channel: adjustment of, 220–9; capacity, 155; deposits in, 194; deserted, 60; diversion of, 202–11; flow, 120; inflow to, 139; pattern, 197–235; roughness, 225; secondary, 26; storage, 120; straight, 211–17, 252–261
chronology of underfit streams, 270–2
circulation, subterranean, 99
climate, 98–100, 110; change of, 21, 33, 88, 219, 265
cloudburst flood, 150
coarseness of terrace deposits, 40
Coastal Plateau, 73
coefficient: of determination, 259; of friction, 101
coherence, 96
comminution, 97, 105

Index

compartments, compartmented valleys, 30–1, 55
competence, 104, 239
components of flow, 160
composition: of drainage net, 122–3; of stream systems, 127
compound fluid, 102–3
concavity, 239
concentration: of suspended sediment, 186–7; of sheet flow, 154; of water, 104
cone, alluvial, 27
cone-in-cone, 28
configuration of channel, 233
confluence: existing, 93; former height of, 91; former level of, 93; preglacial, 93
consequent drainage, consequent system, 115–16
continuum of natural stream channels, 235
controls, geological, 139
converging surface, 112
convexity, 113
corrasion, 61, 95–7, 107–15; and declivity, 105–7; and transportation, 105
course, constrained, 56
critical distance, 154, 156
critical length, 147–9, 161–2
cross-currents, 101
cross-grading, 155–9, 161–2
cross-section, 210, 228, 255; form of, 103; size of, 103
crumb structure, 140
currents, subsidiary, 100
curvature, 39 ff., 253; radius of, 39
curve: calculated, 85; circular, 250; smooth, 90
cusp, 38, 46, 69; blunt, 50–1; defended, 49–50, 52–6, 63, 67, 69–70; four-swing, 48; free, 50, 56, 63, 70; one-sweep, 47–8, 51, 65; sharp, 51–2; three-swing, 48; two-sweep, 47, 51, 59, 65; two-swing, 47
cusp-making, 54
cut-off, 23, 43–4, 70–1
cutting, lateral, 49
cycle, 28, 32; of erosion, 75

Darcy–Weisbach coefficient, factor, 211, 222–3, 233, 257
datum line, inclined, 80

debris: comminuted, 98; heterogeneous, 102
declivity, 97–9, 104, 110, 113; and volume, 107–8; of branches, 107; of main stream, 107. *See also* gradient, slope
decrease: of concentration, 187; of stream volume, 40–1, 43, 61–2
deepening, 201
definition of flood plain, 166
deflection, 57; of flow, 203
deflection pools, 22–3
degrading, degradation, 43, 46, 49, 54, 57, 64, 68, 71, 95, 108, 192–3, 202, 207, 222, 227, 231
delta, 19, 27; terrace, 20, 27–30; terrace, lateral, 27
denudation, 88, 108
deposit: lateral, 176; within-channel, 176
deposition, 24, 44, 145–6, 233–4; on agricultural land, 182; on the convex bank, 178; within the channel, 177
depositional front, 225
depression storage, 142, 144
depth, 142, 200, 202, 208, 221, 225, 229, 231, 234, 239, 255; mean, 208; of fill, 190–1; of flow over bar, 177; of overland flow, 146; of surface detention, 139, 142; variance of, 258–261
depth/sinuosity characteristics, 270
desert regions, 154
desiccation, 272
determination of physiographic factors, 129–33
development of stream systems, 154–161
dimension, ultimate, 155
discharge, 33, 105, 202, 215, 221, 227, 229, 233; channel-forming, 272–3; dominant, 213–14; effective, 222, 231–2; mean annual, 190
discharge ratio, 273
discontinuity of gradient, 73, 80–1
discordance, 79; of level, 41
disintegration, 95, 98
distortion, 23
distribution: of drainage lines, 109–10; of rainfall, 98–9; of underfit streams, 266
distributions, geological, 109–10

Index

diversion, 158, 265
diversity of particle size, 182–3
divides, interior, 160
dominant flow, 191–2
downstream shift, 23
drainage: area, 88, 129; composition, 133; density, 118–20, 122, 127, 133, 156; density, computed, 133; inconsequent, 115–16; lines, secondary, 111; net, dendritic, 155; pattern, 122, 136; slope, 113
drainage-basin topography, 156–62
drift, 37–8; fill, 71
Du Boys formula, 147
dunes, 59, 225, 254
dynamic equilibrium, 110

earth slips, 158
eddies, 151–3, 172–3, 176–7, 232
elevation, 32, 41, 115; of coast, 33; of point bar, 176
emergence, 19
energy, 100–3, 105, 153; dissipation of, 253; eddy, 153; expenditure of, 101, 254–5; kinetic, 153; loss of, 253; rotational, 153
energy grade line, 223
entrainment, 146
envelope profiles, 80
ephemeral streams, 119
equality of action, 110–11
equilibrium, 28; dynamic, 110; of action, 106–7
eroding force, 146–7
erosion, 108; accelerated, 184; concave bank, of, 178; control of, 97 ff.; lateral, 186; longitudinal, 160
erosional processes, 95 ff.
exception to law of divides, 113
extrapolation, 79, 81–2, 84, 90

fall, 102, 106
fan, 39, 60; terrace, 27–8, 30
Fanog base-level, 75, 91, 93
Fanog period of base-levelling, 94
flood, 26, 78, 99–100, 106–7, 155, 192, 194, 220; annual, 166, 171; cloudburst, 150; damage stage, 169; definition of, 168–9; hurricane, 179; inundating, 168–9; most probable annual, 272; record, 179; stage, 106–107, 169
flood plain, 40, 42, 44, 61, 64, 67–9, 71, 166–94; aggrading stream of, 192–3; deposits, 171 ff.; mean height of, 169–71; stable stream, of, 189–92
flow: laminar, 142, 153; mixed, 144; non-turbulent, 142; turbulent, 153, 162
flume experiments, 202–11, 214–16, 221–2, 225–9
force of raindrops, 99
form of bed, 101
frequency: of flooding, 192–4; of flood-damage stage, 169–71; of overbank flow, 166–71, 184
friction, 100, 102–3, 146; factor, 255–7; flow of, 101. *See also* Darcy–Weisbach coefficient
frost, 96, 98–9
frozen ground, 266

geometrical: progression, 123; series, 123
geometry of meanders, 239–52, 261
glacial epoch, 87
glacial erosion, 77, 89
glacis slopes, 27
glacis terrace, 21, 25–6, 38
gorge terrace, 20, 29–30
gradation, 154; of divides, 156–61
grade, 43, 50, 75, 106; inequality of, 106
gradient, 34, 74–5, 77, 80, 82, 84–6, 88, 90, 93, 109; average, 76; change in, 76. *See also* declivity, slope
gradient curve, 81, 91
grain size, 221–4, 226, 229, 231; vertical gradation in, 182. *See also* calibre
grass, grass sod, 151–2
gravel banks, 33
gravity, 95–7
ground slope, 120
gulches, 39
gullies, 28, 154

hanging valleys, 89–91
head of rejuvenation, 84, 89
height of ancient base-level, 79
High Plateau of Central Wales, 73
horseshoe bends, looping, 23, 242
Huttonian theory, 19
hydraulic: factors, 233, 235; gradient, 140; head, 140; radius, 142; resistance, 239; variables, 239

Index

hydrograph, 230
hyperbolic function, 80

incised meanders, 246
increment of sediment, 185
index of turbulence, 144–5
inequality of grade, 106
infiltration capacity, 119, 139–41, 145, 150, 154, 162
infiltration theory of surface run-off, 139–45
ingrown meandering valleys, 267
ingrown meanders, 264–6
initial resistivity, 119
insufficiency of supply, 104
intensity: of rainfall, 140, 145; of surface run-off, 141–2
interdependence, 110–11; of slopes, 114
interfluve hills, 158–60
interscarp space, 42–3
intersection of slopes, 112
island, island formation, 201, 203–5, 211, 217
island areas, 205

jostling, 97
junction, concordant, 161
junction terrace, 27

kame, 59

lake: ice-dammed, 40; pluvial, 271, 274
lamination, 182
land sculpture, 108
Llandovery base-level, 75, 91, 94
late-mature stage, 94
lateral delta terrace, 27
lateral terrace, 20, 25 ff.
law: of declivities, 108–9; of divides, 109–14; of divides, exception to, 113; of equal declivities, 114; of overland flow, 141–3; of stream lengths, 126; of stream numbers, 126; of stream slopes, 129; of structure, 108–11; of uniform slope, 108. *See also* Playfair's law
laws of drainage composition, 123 ff.
ledges, 49–51, 54, 62–4, 69–70
length, 123 ff.; of overland flow, 120–2, 145; of sheet flow, 120; of streams of first order, 132
lenses, 182

levees, natural, 183–4
level: of old valley, 82; of valley floor, calculated, 82–4
limiting volume of eroded material, 153
limits of error, 82
line of maximum depth, 211
load, 42–3, 79, 87–8, 104, 203, 218, 225, 227–30, 232, 234; excess of, 105
lobate spur, 70
lobe, 44
logarithmic: component, 80; curve, 80; function, 80
loess, 271–2
long profile, longitudinal profile, 73 ff., 90
lowering of base-level, 94
Lower Plateau, 73

magnitude–area–intensity analysis of precipitation, 273–4
manifest underfit, 265, 267
Manning formula, 141
mantle, 88
master river, 71
mature: floor, 90; profile, 77; stage, 77
meander, 217, 235; compressed, 53, 57, 64; enclosed, 44; enlargement of, 44; geometry of, 239–52, 261; incised, 246; ingrown, 264–6; length of, 215–252; slipped, slipping, 50, 64
meander belt, 43–5, 55
meander trough, 269
meandering: channel, 211, 214; valley, 264–74; valley, ingrown, 267, 270
meltwater, 266; floods, 21
mesh length, 126, 133
micropiracy, 155–6, 160
migration, lateral, 194
minimum variance, 239
moisture content, 139
moraine, 59, 271
mountain track, 30

Nant Stalwyn base-level, 75, 91, 93–4
node, 55–6; fixed, 56, 68; free, 56, 62
noise, 252
number, 123 ff.

open system, 239
order, 123 ff.; assumed, 130–2; determination of, for main stream, 132; inverse, 130–2

outwash, 59
overbank deposition, deposits, 171, 178–9, 182–9, 194; amount of, 186–9
overbank flow, 176; frequency of, 166–171, 184
overflow, 177
overland flow, 141 ff., 232; profile of, 143–5; surface erosion by, 145–53; types of, 145
overspill, 265
overtopping, 155–6

parabola, 250
parabolic function, 80
particles, 231
path of greatest probability, 240–1
pattern, straight, 235
percolation, 99, 266
period of base-levelling, 77, 90
periodicity, 33
perennial streams, 118–19
piracy, 265
plain track, 30
planation, 23–4, 30, 33
plant cover, see vegetation
Playfair's law, 127, 129, 155, 161
plot experiments, 142–3
pluvial lakes, 271, 274
pluviosity, 271, 274
point: of convergence, 112; of rejuvenation, 77, 91
point bar, 171 ff., 194
pollen, fossil, 271
pollen zones, 271
pool and riffle, 213–17, 253–5, 268
pools, 266
position of divide, 112
post-glacial erosion, 87
post-glacial period, 34
pounding, 97
power, 239; to transport, 97, 99
power function, 216
pre-glacial profile, 77
pre-glacial valley, 79
principle of equal action, 112
profile, 109, 213, 239, 253; ancient valley of, 79; average, 76; calculated, 82, 85; concave, 112–13; convex, 112–113; drainage slope of, 112; mountain, 109; overland flow of, 143–5; smoothed, 82, 89; transverse, 75
proglacial lakes, 272
pulsation, 32

quantity of detritus, 97
quasi-equilibrium, 211, 218, 220, 222, 235

radiocarbon, see carbon-14
radius of curvature, 242–3
rain, rainfall, rainstorms, 78, 88, 95–100, 104, 115, 119, 162
rain intensity, 145; and erosion, 149–51
raindrops, force of, 99
rainfall distribution, 98–9
rainfall, excess, 141
rain-impact erosion, 158
random walk, 240–1, 251
range or particle size, 222
rate: of change of infiltration capacity, 140; of corrasion, 97, 106; of cutting, 44; of degradation, 100; of disintegration, 100; of downcutting, 42; of elevation, 20; of energy expenditure, 261; of energy loss, 261; of erosion, 97 ff., 110; of lateral migration, 178, 180–1; of migration, 186; of sand feeding, 202; of surface runoff, 139; of swing, 44; of transportation, 100, 104
ratio: of erosive action, 109; of resistances, 109; w/w, 267
ratios: of depths, 210; of slopes, 210; of widths, 210
rectangular hyperbola, 80–1, 91
recurrence interval, 166–70, 179, 192; of overbank flooding, 168
reef, 50
re-entrants, 57, 63–4, 67–8
rejuvenation head, 75, 90
relation: of declivity to transportation, 102–3; of drainage area to stream order, 127; of geological structures to drainage composition, 133–9
relief, 119; on flood plain, 186
resistance, 106, 222–8, 231, 233
resistivity, 154; to erosion, 145–9
resultant slope, 120, 160
retardation, 110
Reynolds number, 222
ridges, secondary, 111
rill, rill channel, 104, 109, 141, 154–6, 232
rill stage, 154
rilled surface, 154
ripples, 225
river, rockbound, 31

rock: barrier, 55, 66, 76; channel, 76; floor, 89; node, 68; step, 79, 89–91; texture, 98
roughness, 187, 222–5, 230–2, 234; factor of, 141, 144, 148; relative, 153, 222
rubbing, 97
run-off, 78, 88, 139; exceptional, 232; intensity of, 147–8

saltation, 151–2
sand, 96; transport of, 189
sandstone barrier, 65
sapping, 96, 98
scar, 33
scarp: ascending, 38; one-sweep, 47
scour, 179, 192, 205; local, 182; maximum depth of, 191–2; potential depth of, 191; probabilities of, 192
scroll, 44
sculpture, 95, 108–14; and declivity, 108
sea-level, 20
sediment: concentration, 194, 203, 221; maximum, 187; point bar of, 172–5; storage, 178; transport, 153, 225–8
sedimentation, 151–3
shear, 225–8, 230–31; velocity, 226
sheet flow, *see* overland flow
shift, shifting, 23, 44; of bars, 205; of channel, lateral, 172; of divides, 114; of waterways, 114
side-slope erosion, 162
silt, transport of, 189
sine curve, 213–14, 244–5, 247, 250
sine-generated curve, 241 ff., 261
sinuosity, 211, 217, 248, 250–2, 258, 270
sinusoidal function, 242
size of fragments, 97. See also calibre, grain size
size limit of detritus, 101
slope, 39, 41–3, 61, 120–2, 144–5, 147, 187, 202, 207–8, 217–19, 221, 225, 227, 229–31, 234, 239, 255; angle, 146; change of load, related to, 228–229; energy profile of, 257; mean, of channel bed, 224; pregraded, 161; radial, 27–8; ratio, 120; series, 110; water surface, 223. *See also* declivity, gradient
slopes, 20, 77; convex, 77
slough, 199
smoothness, relative, 223

snowfall, 40
snowmelt run-off, 253
soil, 99–100; erosion, processes of, 145–6; structure, 139; surface, condition of, 139; texture, 139
solution, 95–7, 99, 157
sorting, 203–4, 207; action, 24; local, 203
spacing of riffles, 268
spillwater, 266
spurs, 44; buried, 64; defended, 62; free, 62
steady state, 239, 260
steepening, 206–7
steepness, 254
stillstand, 28–9
storminess, 274
storms, 155–6
stream: composition, 136; frequency, 122; fully loaded, 105; gradation, 161–2; gradient, 74; intermittent, 118–19; length, average, 130; length ratio, 129–33; order, 117–19
streamline pattern, 173
streamlines, 201
structure, 115; biological, 139; geological, 134
subglacial streams, 40
submergence, 21
subsidence, 29
superimposition: by alluviation, 115–116; by planation, 116; by sedimentation, 115
superposition, 49, 55
of curves, 85–7
supply: of debris, 106; rate of, 141
surface: contact, of, 103; detention, 120, 139, 142, 145, 150, 157; roughness, 145, 147; run-off, *en masse*, 150; run-off intensity, *see* intensity of surface run-off
suspended load, 151
suspended sediment, 186–7
suspension, 97
sweep, sweeping, 23, 45–6, 55, 57, 60
swing, swinging, 45–6, 50, 53, 55, 57, 60, 65, 68–9, 71; breadth of, 44; diminished, 53–4; lateral, 43, 186
system: of displacements, 115; of stream orders, 117

talus fans, 27
tearing loose of soil material, 145

Index

temperature, 98–9; change of, 85–97
terrace: amphitheatre, 25–6, 31–2; coastal, 19; defended, 71; delta, 20, 27–30; delta, lateral, 27; fan, 27–8, 30; glacis, 21, 25–6, 38; gorge, 20, 29–30; junction, 27; keeping, 54; lateral, 20, 25 ff.; pattern, 36; period, 34; scarps, 36; sequence, 179; spur, 38; trailing, 50; undefended, 55
terraces, origins of, 22
terracing, miniature, 32–3
Tertiary era, 88
texture, 122; of rock, 110
thalweg, 211, 249
tilting, 87, 89, 94
torrential period, 33
tortuosity, 23
total length of streams, 126
transmission capacity, 140
transport, 146; of eroded material, 145
transportation, 95–7, 99, 151–3; and comminution, 100–2; and declivity, 102–3; and volume, 103–5
transporting power, 24
travelling of river curves, 23–4
Triassic, 58
tributaries, fingertip, 122; unbranched, 117
turbulent flow, 153, 162

unconformity, 115; topographic, 74
underfit stream, 264–74; manifest, 265, 267; of Osage type, 268, 270
uniform grade, 106
unstable fluid, 102–3
upheaval, uplift, 28–30, 32, 36, 43, 49, 57, 90, 94; postglacial, 41

valley: ancient, 87; cross-section of, 162; floor, ancient, 85; gradation, 161–2; pre-existing, 77; side-slopes of, 74; track, 30
valley meander, ingrown, 267, 270
values: for former tributary junctions, 90; of constants, 81–4, 91–3; of variables, 124

variables, 124, 217–18, 234
variability, 261
variance, variances, 249–50, 252, 258–261; minimum, 239; of bed shear, 258; of depth, 258–61; of slope, 258–261; of velocity, 258–61
variation in morphological features, 139
vegetation, 88, 95–7, 99, 104, 110, 139, 145, 189, 199, 201–2, 220; and transportation, 103–4
velocity, 20, 24, 97, 99–107, 153, 172, 186, 189, 194, 200, 210–11, 221, 225, 229–32, 234, 239; distribution, 153; inequality of, 106; of fully loaded stream, 102; of overbank sections, 187; of overland flow, 144, 146; profile, 226; of turbulent hydraulic flow, 141–2; variance of, 258–61
viscosity, 101, 103
volume, 39–40, 42–3, 49, 78–9, 86–7, 99, 104, 107–10; change of, 64; decrease in, 40–1, 43, 61–2
vortex: motion, 153; ring, 152

wandering, 43–6, 49–50, 55, 68; of thalweg, 211–13, 220
waste, 110
water surface: profile, slope of, 224, 253, 255, 257
watershed line, 158
wave-length, 210–17, 250; ratio, 273
wave-length/width ratio, 267–8
waves: of run-off, 151; of translation, 19
wear, mechanical, 96
weathering, 95–6
weight of particle, 101
widening, 201
width, 199, 202–3, 208, 215–16, 220–2, 229, 231–4, 239
width/depth ratio, 219
winding streams, 22
winnowing, 227, 234

zero azimuth, 243